KU-339-275

SKETCHES FOR
THE FLORA

Other books by W. Keble Martin

THE CONCISE BRITISH FLORA IN COLOUR

OVER THE HILLS (an autobiography)

SKETCHES FOR THE FLORA

W. KEBLE MARTIN

Selected and with introductory text
by John Caldwell

Foreword by Wilfrid Blunt

London
MICHAEL JOSEPH

First published in Great Britain by MICHAEL JOSEPH LTD
52 Bedford Square, London WC1B 3EF
1972

7181 0945 7

Photoset and printed in Great Britain by BAS Printers Limited, Wallop, Hampshire in Ehrhardt eleven on thirteen point on paper supplied by Frank Grunfeld Ltd and bound by Dorstel Press, Harlow

Foreword

SOME twenty or so years ago, soon after I had published my *Art of Botanical Illustration*, a colleague of mine at Eton asked if he might bring a friend of his to see me—an elderly Devonshire clergyman who had painted some pictures of wild flowers. I entertained no great hope that I was about to be shown anything more than some rather amateurish sketches of the flowers that happened to grow in and around his parish; I therefore was amazed when he produced a virtually complete British Flora executed with great skill, deep botanical knowledge and almost unbelievable attention to detail.

He was anxious to publish them; but such was their delicacy that I doubted whether any process of reproduction at that time available could do them justice at a price which would make the proposition commercially acceptable. Fortunately, however, improved methods of colour-printing at relatively low cost were in due course developed, and in 1965 what was to prove the year's best-seller made its appearance: The Rev. W. Keble Martin's *Concise British Flora in Colour*. The interest it aroused encouraged its author to follow it up with an autobiography of quiet charm—the sort of book to read at ease in a deck-chair in a garden full of Madonna lilies and delphiniums.

It is always exciting to see an artist at work informally; it was therefore a happy idea of Messrs Michael Joseph to propose to Keble Martin's widow that they should be allowed to publish a volume of field sketches from which the highly finished paintings had been made. It would be idle to pretend that these sketches have the swiftness of line of those of an internationally great botanical artist such as Ehret; they have, however, a precision and a clarity which give them distinction, and it is also interesting to compare them with the finished products—to observe those little modifications to leaf or stem or flower which Keble Martin had seen fit to make in order to achieve his orderly compositions. In a few cases where important original drawings have been lost, he has himself made tracings from the finished paintings to fill the gaps. On plate 28 of this book there is a drawing of *Rubus villicaulis* for which, as will be seen from the accompanying note, there was no room in the *Flora*.

Herbals and florilegia of the sixteenth and early seventeenth centuries were often intended to be coloured by the purchaser, and sometimes an indication was given of the colours to be used. Many of those recommended by, for example, Crispin vande Pas for his *Hortus Floridus* (1614) may sound strange to modern ears, and modern paint-boxes are unlikely to contain masticott, verdigreece or maydens blush. The drawings in Keble Martin's book have been printed on a paper suitable to take water-colour, and owners of it who live in the country may find it an agreeable pastime to follow the example of an earlier age by painting in each flower as they find it.

Wilfrid Blunt

Introduction

DR. W. Keble Martin had collected in a number of loose-leaf books most of the sketches of plants he had drawn over a period of some sixty years, in preparation for the coloured illustrations in the volume which was ultimately published in 1965 as *The Concise British Flora in Colour*.

It was suggested to his widow, Mrs. Flora Martin, that many of those who have copies of the *Concise Flora* would be interested in seeing a selection of the sketches from which the illustrations were finally prepared. This volume contains a short account of his life and a number of the sketches drawn at different times which were ultimately used in the preparation of the *Flora*.

For over two thousand years from the time of the Greeks, Botany was closely associated with the practice of Medicine. This is not surprising when one realises that plants were the main source of drugs and that the pharmacology of the times was dependent on a knowledge of plant products. In the middle ages the Chair of Botany in the teaching institutions was always associated with the Chair of Medicine and of Anatomy. Nor has the connection been completely severed at the present time, as in some of our older Universities even today the Professor of Botany is still a member of the Faculty of Medicine—a somewhat curious survival of an earlier tradition.

The first Professor of Botany who was not a physician was Ghini who, in the 17th Century, occupied the Chair of Botany in the University of Bologna.

In more recent times Botany seems to have been more commonly associated with Theology and it is surprising how many clerics have been botanists of one kind or another. The names of such people as Stephen Hales of Teddington and Gilbert White of Selborne and many others come immediately to mind. A considerable number of academic botanists in the last two hundred years have taken Holy Orders but sometimes this may have been more associated with the Fellowship requirements in the older English Universities than with any intention on the part of the ordinands to make a career in the Church.

It is the more interesting, therefore, to find that a man like W. Keble Martin, a graduate of the University of Oxford in Greek Philosophy and Botany, who spent the whole of his long and active life as a hard-working and faithful parish priest, was at the same time a botanist of distinction. Towards the end of his life he achieved considerable fame with the production of *The Concise British Flora in Colour* which became the best-seller of the publishing season in 1965/66. The Annual Report of the Department of Botany of the University of Oxford for 1965 claimed Keble Martin as one of the three best-known past students of the Department, linking his name with that of Sir Joseph Banks, P.R.S. (1743–1820) the distinguished traveller and naturalist, and that of Sir John Lawes, F.R.S. (1814–1900) the founder of Rothamsted Experimental Station.

Although most people associate Keble Martin with his very successful *Concise British Flora* which, as we have seen, established his reputation when he was approaching his 90th year, it is probably less well known that he was responsible, with the late Gordon T. Frazer and others, for the monumental *Flora of Devon* which was published in 1935 and which was so successful that it has long been out-of-print. This was one of the first, if not the first, of the modern County Floras and was a model for many of those which followed. Incidentally, Keble Martin was keenly interested in the preparation of a new County Flora for Devon which will treat the whole subject from a more ecological aspect, in accordance with present-day practice.

The fact that he was so interested in this project when well over 90 years of age indicates the depth and width of his interest in the subject. He was a member, until his death, of the Editorial Committee of the Botanical Section of the Devonshire Association and was most faithful in his attendance at the meetings of the Section. He very regularly brought to them specimens of some rare or unusual plants which he had found in the course of his regular botanical excursions.

In 1968 Keble Martin published his autobiography under the title of *Over the Hills*. This gives an indication of his energetic pursuit of plants which he endeavoured to find in their native habitats and which he studied *in situ* often bringing a specimen back to ensure that when he made his initial illustration he had got the exact colour of the flower and all the details correct. He records that on one occasion he brought a specimen of a rare saxifrage from Ben Lawes, drew it, and then climbed back the next day to replant it in the same place. It is fortunate that he kept almost all his original sketches—some 1800 in all, which he gathered into five books. The present volume contains some of the more interesting sketches which were used to make up the plates of the coloured illustrations in the *Concise Flora*. The whole work, as has been noted, covered a period of over sixty years and involved a great deal of travelling as well as requiring a considerable degree of artistic ability.

Keble Martin makes it clear in his autobiography that he found the wild flowers of the British Isles a fascinating study to which he devoted so much of his scant leisure time. It became an absorbing hobby which enabled him to carry out better the exacting duties of a hard-working parson. He tells us that his grandfather was Rector of Staverton, a small village on the river Dart in South Devon. His father, after having been a Scholar at Winchester, went on to New College, Oxford. After ordination and his marriage to the daughter of his old Headmaster who was subsequently Bishop of Salisbury, he became Vicar of Wood Norton-with-Swanton near East Dereham, in Norfolk in 1879 when Keble Martin was two years old. A few years later the family moved to near Devizes where a visiting uncle, Edward Moberly, introduced the boys of the family to an interest in butterflies and moths. Keble Martin made and kept a very large collection of moths and butterflies of which he was justly proud, and in which he retained an interest throughout his whole life. When he was about 14 years old the family moved to the village of Dartington in Devon. All his young life he was, therefore, associated with the countryside and with rural pursuits which developed his interest in natural history. At Marlborough, also, an active Natural History Society fostered his interest in the subject.

In 1896 Keble Martin went up to Christ Church and read for a Pass Degree in Greek Philosophy and Botany—taking the former subject out of deference to his father who had lectured on it at that College. S. H. Vines was the Professor of Botany at that time and A. W. Church was the Lecturer in the Department. Church was a remarkable man with a wide interest in the subject and as a most competent floral artist may well have encouraged Keble Martin in the sketching of flowers which later became one of his major interests. At Oxford, also, Keble Martin was fortunate in making the acquaintance of G. C. Druce, who stimulated his interest in systematic Botany and who, in later years, sent him many specimens from all parts of Britain.

Incidentally, Keble Martin reports in his autobiography that the first of the drawings which ultimately appeared in the *Concise Flora* was made in Dartington in 1899. The first sketch he made was of snowdrop flowers and ivy leaves (see plate 83). He decided in his later sketches to use only the foliage of the plants of which he drew the flowers.

After training for the Ministry at Cuddesdon Theological College Keble Martin was ordained deacon in 1902 in Southwell Minster and then became a curate in Beeston a suburb of Nottingham. From Beeston Keble Martin went to Ashbourne in Derbyshire as curate with Canon Morris. He mentions in his book that on one occasion while at Ashbourne he cycled the long distance to Lincoln to visit Edward King the distinguished and revered Bishop of that diocese.

While at Ashbourne Keble Martin became engaged to one of the parishioners and feeling he should find another curacy while he was waiting for a living of his own, he accepted the senior curacy at Lancaster Parish Church. Up to this time he had been regularly, if somewhat haphazardly, making sketches of the more unusual plants which came his way. About the time of the move to Lancaster, however, he obtained a copy of the *London Catalogue of British Plants* and began to organise his sketches into the form which ultimately led to the production of the plates subsequently published in the *Concise Flora*. This was very much in the future, however, and the death of the Vicar of Lancaster left a great deal of the work of a large parish in Keble Martin's hands.

In 1909 Keble Martin was appointed Vicar of Wath-on-Dearne and in the same year he married. The vicarage had been condemned and it was necessary for the parish to raise funds for the building of a new one. Furthermore, there was a daughter parish and a marked shortage of buildings for the work of the Church in all its branches, so the vicar had the further responsibility of raising funds and organising the building of suitable premises in addition to the normal work of a parish priest.

While Keble Martin was at Wath the first World War broke out and a great deal of additional work fell to him to carry out. To add to the difficulties, the parish church was badly damaged by fire. In addition, to help with the work of the Church overseas, Keble Martin volunteered to serve as an Army Chaplain and went to France as Chaplain with the Northumberland Fusiliers.

On returning to Wath after the Armistice Keble Martin worked on the Deeds connected with the parish and showed his interest in this field of study by publishing a history of the area.

For various family reasons Keble Martin, in 1920, accepted the living of Haccombe with Coffinswell, near the sea in Devon, which made a complete change from the conditions in the industrial North. As incumbent of Haccombe he was one of the few Arch-Priests in the country. With less arduous duties as a clergyman he was able to turn again to his interest in floral illustration, and many of the sketches which formed the basis of the *Flora* were drawn during the dozen years he was in this parish. From the notes on many of the sketches, it appears that a great many were done in the middle twenties to middle thirties of this century. The greatest care was taken to ensure the maximum of accuracy in the sketches and with a considerable number of them there are notes indicating that they were re-drawn on various occasions when, presumably, better specimens of the rarer plants became available. The notes also indicate that many of the sketches were not only re-drawn but were also re-painted, presumably again when better material became available.

Keble Martin also tells in his autobiography how, on one occasion, while at Coffinswell, he took a train one Sunday at midnight and visited Killin and Ben Lawers in Perthshire returning home during the Thursday night. While at Coffinswell Keble Martin, as he had done at Wath, wrote a history of the area which was later published in the Transactions of the Devonshire Association. He was a member of the Association for many years and, as we have seen, was particularly active in the work of the Botanical Section right up to the time of his death.

Keble Martin also records in his autobiography the discussions which culminated in the building of a new church at Milber near Newton Abbot. The design was unusual in the period of the 1930s and led to some controversy, though at a later date it would not have seemed so revolutionary.

In 1934 Keble Martin accepted the living at Great Torrington in North Devon, and left Coffinswell. It was during the time he was Vicar of Great Torrington that he collaborated, as we have already seen, with G. T. Frazer in preparing the detailed *Flora of Devon*, which was one of the best of the County Floras and is now so hard to obtain as to be almost a collector's piece. The fact that the new Flora now in preparation will have a much more ecological approach than the 1935 edition does not in any way detract from the value of the earlier book. The two authors, and their collaborators, must have spent a vast amount of time and a great deal of detailed work in the preparation of that excellent volume. This work was sponsored by the Botanical Section of the Devonshire Association under whose auspices it was published.

The parish work at Torrington obviously occupied a great deal of the time of so devoted a parish priest and during his incumbency the Second World War broke out which added further to the burden Keble Martin carried. However, he pursued his interest in the British Flora as the dates on many of the sketches in his note books testify. Towards the end of the War Keble Martin left Torrington and took the living of Combe-in-Teignhead in South Devon, where he had years before advocated the building of a church of a design somewhat unusual for that period. This church was ultimately completed to his great satisfaction as he had worked tirelessly to get the scheme accepted.

In 1949 Keble Martin retired and went to live in a bungalow he had had built near Gidleigh, on the edge of Dartmoor. The time and effort involved in organising the building of the church at Milber had reduced the time he had been able to devote to his sketches but he had, with his accustomed zeal, associated himself into the activities of a Committee which prepared a long list of areas of natural beauty and scientific interest in the County of Devon. Most of its proposals were utilized later by the Nature Conservancy and many of the areas it recommended have been appropriately designated. The Martins left Gidleigh and settled in Woodbury which had many amenities not so readily available in the more remote area at Gidleigh. While there, and up to the end of his long life, he regularly took services for other clergy and took charge, temporarily, of many vacant incumbencies. In 1963, shortly after going to Woodbury, his first wife died after a long period of illness.

Keble Martin, as we have seen above, during a period of over sixty years had been diligently drawing from life all the plants of the British Flora. This had involved visits to many parts of the country and led to a considerable amount of correspondence with many colleagues both amateur and professional who had sent him specimens of rarer species which he would otherwise have found great difficulty in obtaining. During this time he had re-drawn many of the plants he had previously sketched. All were subsequently drawn again on a series of more than a hundred plates, on which the flowers were coloured with meticulous care.

In 1965 the *Concise British Flora in Colour* was published. There had, for many years, been many efforts by various people to have the plates published much earlier but the cost of reproducing plates in colour had, up till that time, been absolutely prohibitive. However, a newer, cheaper process had recently been developed and in that year the book appeared with a foreword by H.R.H. the Duke of Edinburgh and to the great surprise of the author he found himself in his 89th year, the best-selling author of 1965. Incidentally, the name he had originally in mind for the collection of paintings was apparently *Comparative figures of British Flora* which he subsequently changed to *Flowers of Britain* and then finally to the title under which they were published.

Not only did Keble Martin become a 'best-seller' in that year but he also married for the second time. Mrs. Martin is responsible for the suggestion that the sketches contained in this volume should be published, as she feels that many of those who have enjoyed the *Concise Flora* would be interested in seeing how the collection was compiled. At the same time it was felt by many that the carefully prepared and detailed sketches should be made available to all who are interested in Keble Martin's work.

Though the work had occupied so much of his time over almost sixty years, there was much for the author to do before the volume was finally published and he himself records that he had to re-name many of the species to bring the nomenclature into line with that of the code agreed by the International Botanical Congress.

An honour which gave him great pleasure was the offer from the University of Exeter of the honorary degree of Doctor of Science and this was conferred on him in recognition of his work on the British Flora. The conferment took place at a Congregation on 30th June 1966 when the Chancellor, the Dowager Duchess of Devonshire presided. The next year, in 1967, Keble Martin was invited to paint some stamps for the Postmaster General. He was flattered by this invitation and selected some of the most suitable illustrations from the Flora for the purpose.

So into his ninety-first year Keble Martin had lived a full and active life, combining, as we have seen, his work as a parish priest with his wide and deep interest in natural history. While flowers were his chief interest, he treasured, as I well remember as we stored the cabinet for some years in the Hatherly Biological Laboratories at the University of Exeter, a collection of moths and butterflies which he had made over a long period and which he greatly valued.

He died at his home at Woodbury on November 26th, 1969 and was buried in the churchyard of that village. He had the satisfaction, before he died, of knowing that the work to which he had devoted so much time and energy had at last been published, and had given much pleasure to a wide range of people.

JOHN CALDWELL

On plates 10, 12, 30, 39, 51, 71, 73, 81, 83, 86, 87, 89, 90 and 91 composite groups were redrawn for the *Flora*. Although family names have been given, for reasons of space the names of individual plants have occasionally had to be excluded and can be found by consulting the *Flora* (1969 revised edition); they do not therefore appear in the index of this book. The plants on plates 51 and 86 have been slightly reduced from the originals.

Clematis vitalba
Haccombe Devon
Aug - Oct. 1922 repainted 1924
repainted 1932

Calcareous soils from S. Yorks southwards.

Thalictrum minus L.
Subsp. montanum Wall
Berry Head Devon. June 1924
Partly redrawn from a Babbacombe spec
1932

Limestone Rocks & banks

Clematis vitalba L.
Traveller's Joy, Old Man's Beard

Thalictrum minus L.
Meadow Rue

Anemone nemorosa L.
Wood Anemone

Pulsatilla vulgaris Mill.
Pasque Flower

Ranunculus peltatus Schrank

Ranunculus tripartitus DC.
Little Three-lobed Crowfoot

Ranunculus Baudotii
Slapton Ley, S. Devon V.C. 3
July 28th 1927.

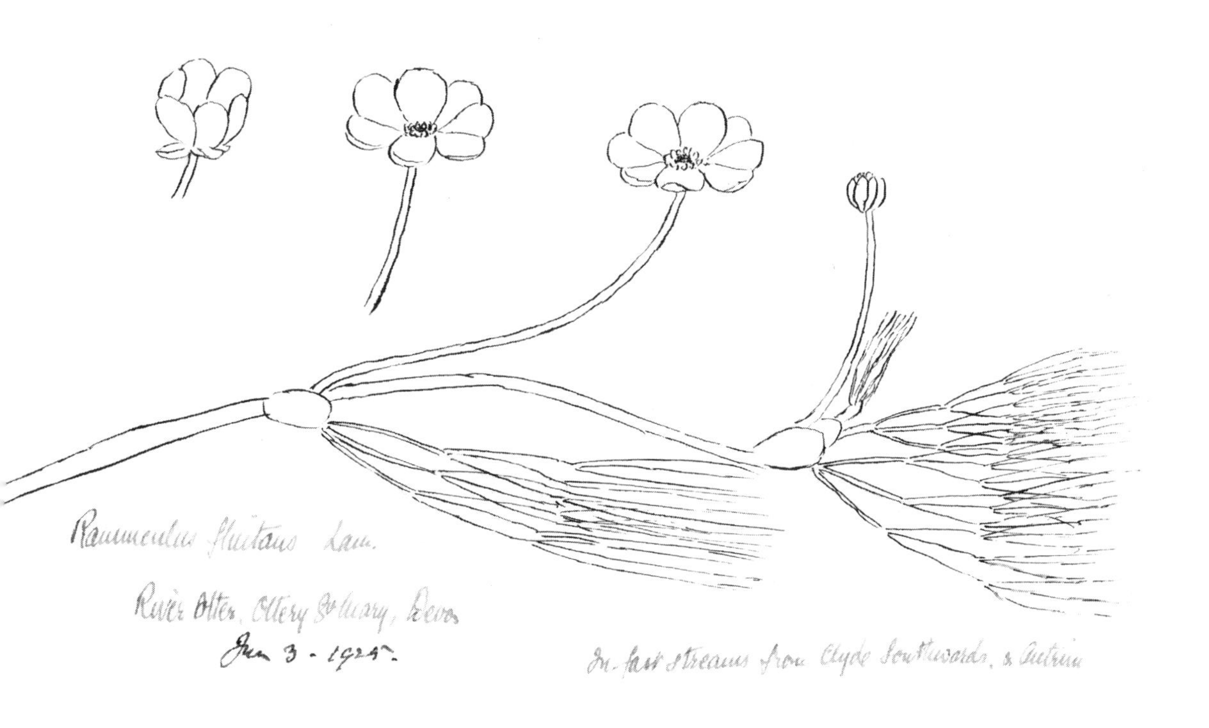

Ranunculus baudotii Godr.
Seaside Crowfoot

Ranunculus fluitans Lam.
River Crowfoot

Plate 3 RANUNCULACEAE

Ranunculus lingua L.
Great Spearwort

Ranunculus ficaria L.
Lesser Celandine

Ranunculus repens L.
Creeping Buttercup

Ranunculus bulbsous L.
Bulbous Buttercup

Plate 4 RANUNCULACEAE

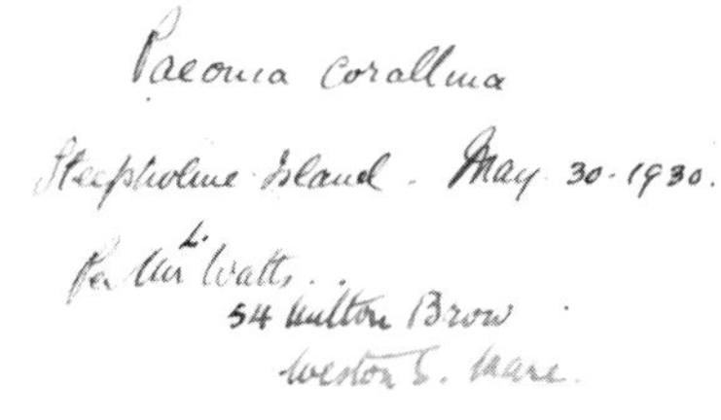

Aquilegia vulgaris.
wood near Holne Dartmoor
transplanted March 20. 1925
flowered & painted May 1926.

Paeonia mascula (L.) Mill.
Paeony

Aquilegia vulgaris L.
Columbine

Caltha palustris L.
Marsh Marigold, Kingcup, Mollyblobs

Helleborus viridis L.
Green Hellebore

Papaver somniferum L.
Opium Poppy

Glaucium flavum Crantz.
Yellow Horned-Poppy

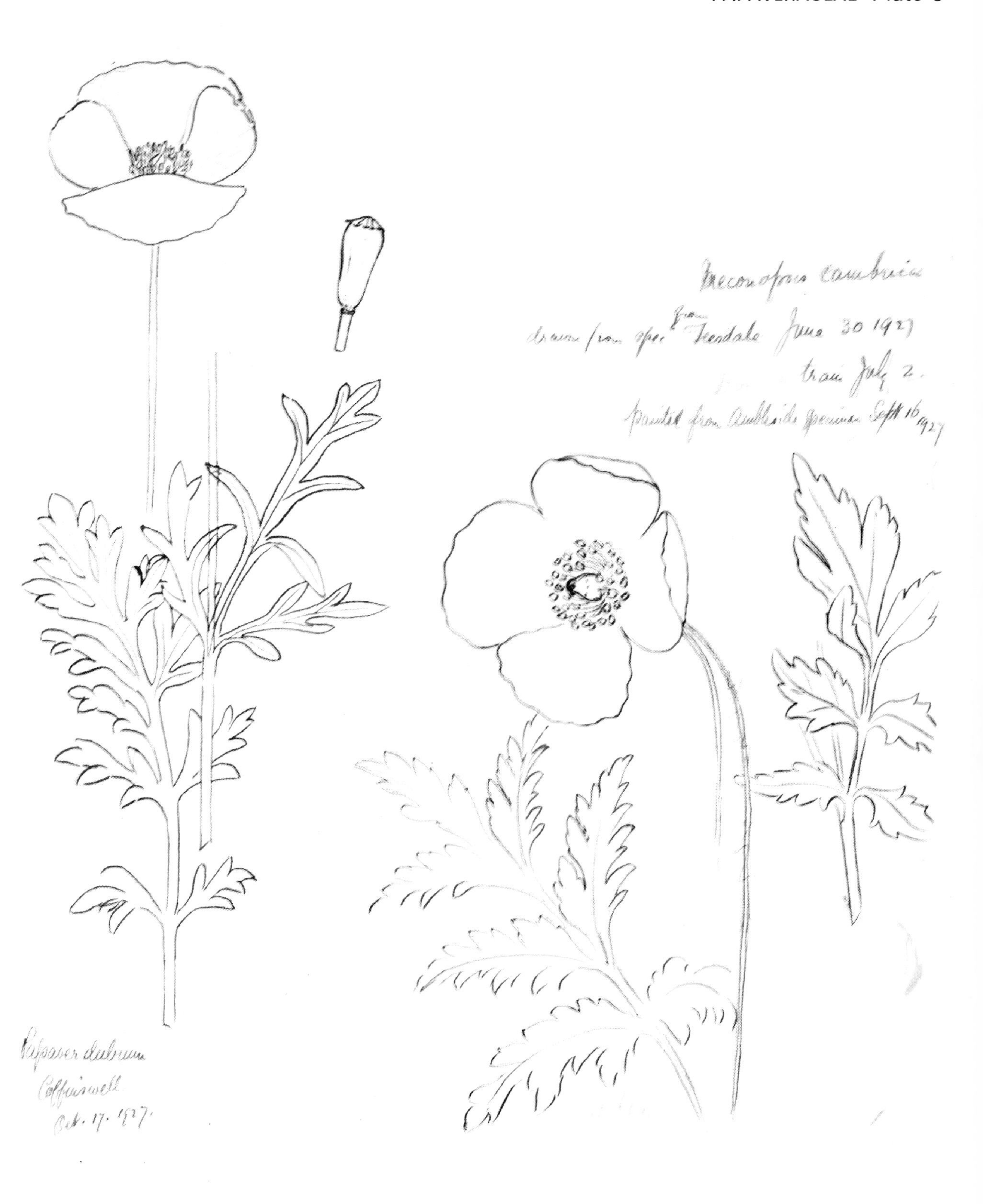

Papaver dubium L.
Long Smooth-headed Poppy

Meconopsis cambrica (L.) Vig.
Welsh Poppy

Plate 6 FUMARIACEAE and CRUCIFERAE

Corydalis lutea

Coffinswell Oct. 24. 1927.

Corydalis lutea (L.) DC.
Yellow Corydalis

Matthiola sinuata (L.) R. Br.
Sea Stock

Arabis glabra (L.) Bernh.
Glabrous Tower-Cress

Barbarea verna (Mill) Aschers
American Cress

Barbarea vulgaris R. Br.
Common Yellow Rocket, Winter Cress

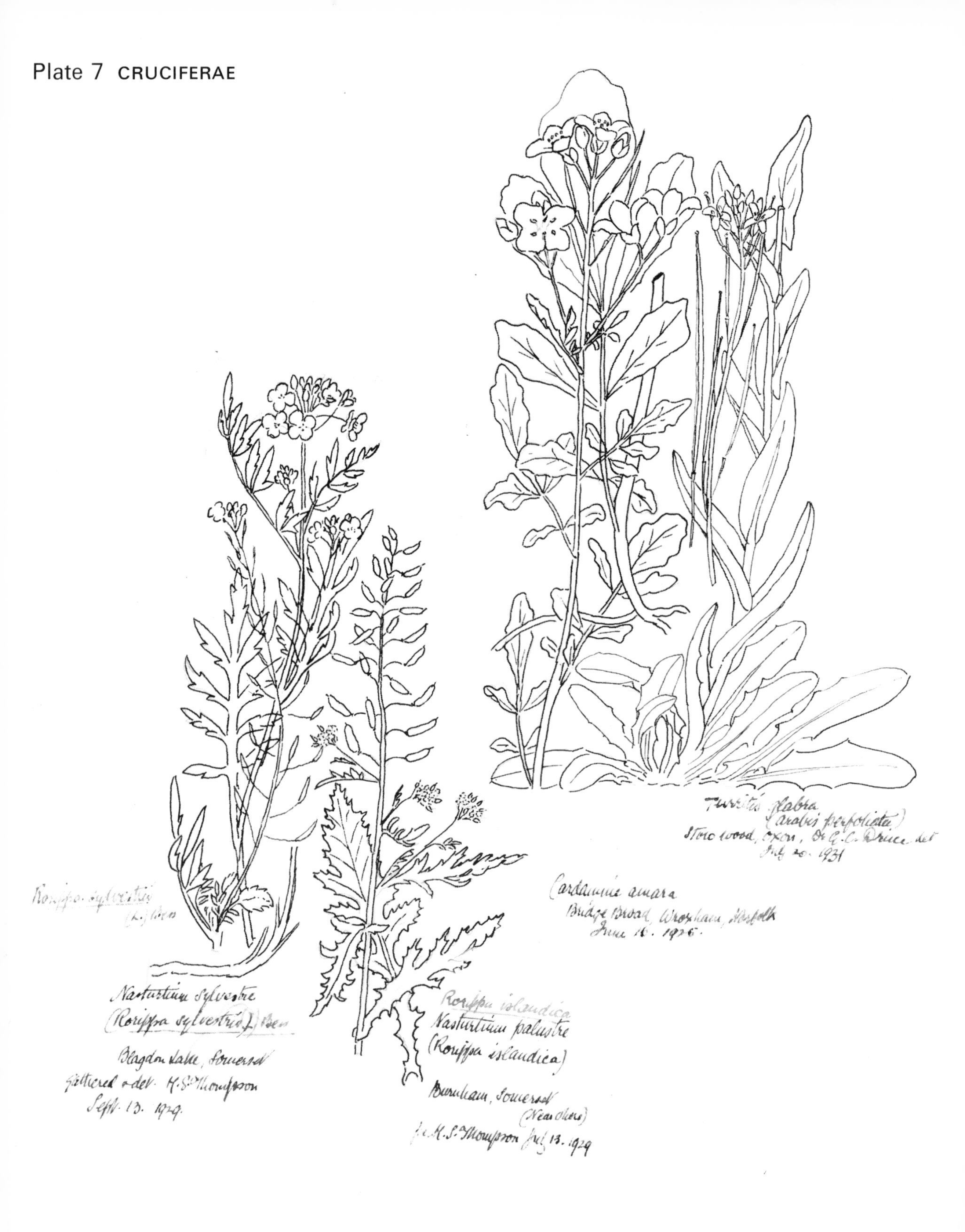

Rorippa amphibia (L.) Bess
Amphibious Yellow Cress

Rorippa islandica
(Oeder ex Murray) Borbas,
Marsh Yellow Cress

Cardamine amara L.
Large Bitter Cress

Arabis glabra (L.) Bernh.
Glabrous Tower-Cress

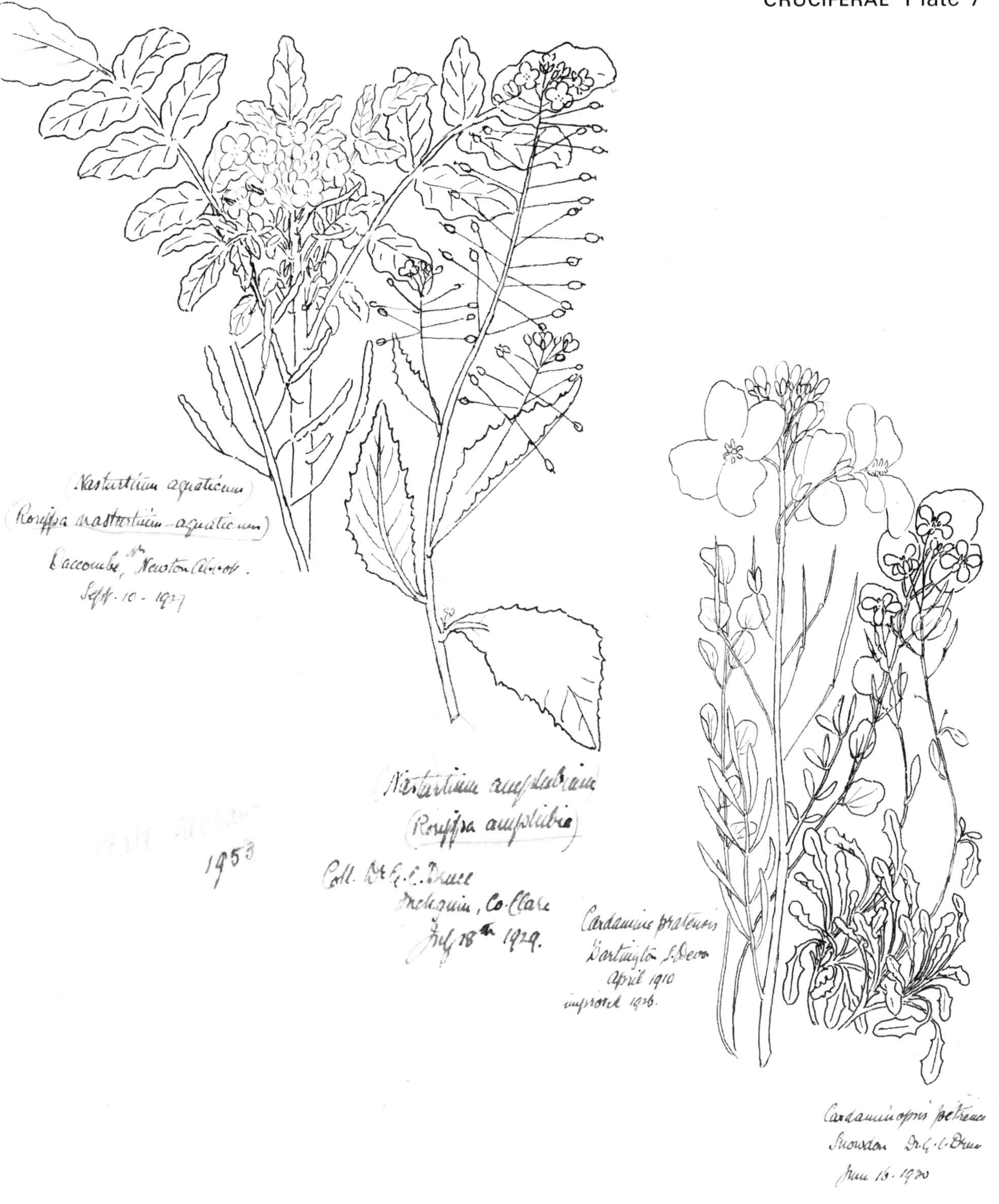

Rorippa sylvestris (L.) Bess.
Creeping Yellow Cress

Nasturtium officinale R. Br.
Water Cress

Cardamine pratensis L.
Cuckoo Flower, Lady's Smock

Cardaminopsis petraea (L.) Hiit.
Mountain Rock-Cress

Cardamine bulbifera (L.) Crantz.
Coral Root

Descurainia sophia (L.) Webb ex Prantl.
Flixweed

Hesperis matronalis L.
Dame's Violet

Cardamine impatiens L.
Narrow-leaved Bitter Cress

Plate 9 CRUCIFERAE

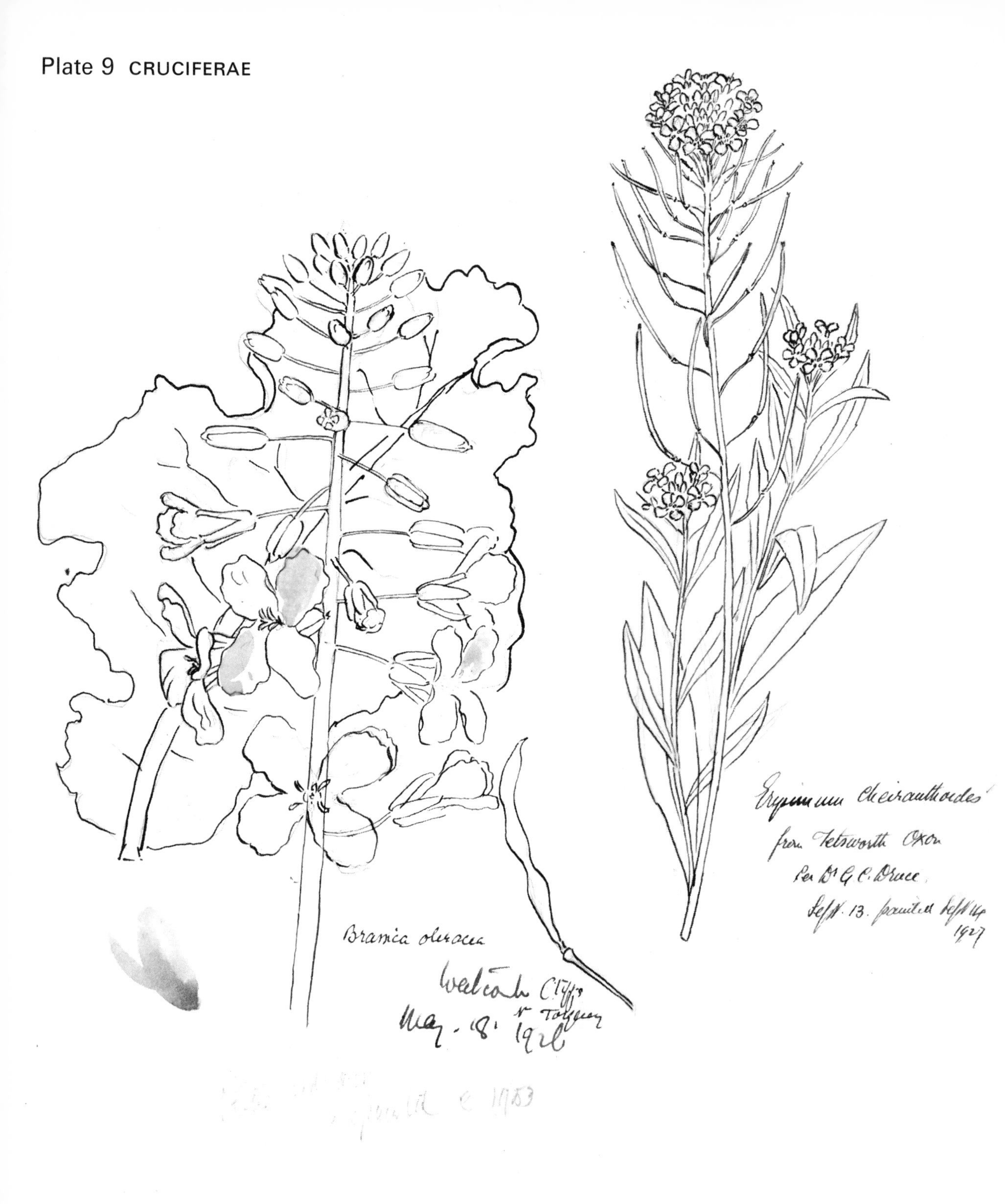

Brassica oleracea L.
Wild Cabbage

Erysimum cheiranthoides L.
Treacle Mustard

Sinapsis arvensis L.
Charlock

Rhynchosinapis monensis (L.) Dandy.
Isle of Man Cabbage

Redrawn from original sketches for the *Flora*

Redrawn from original sketches for the *Flora*

Plate 10 CRUCIFERAE

Iberis amara L.
Candytuft

Lepidium latifolium L.
Dittander

Capsella bursa-pastoris (L.) Medic.
Shepherd's Purse

Thlaspi perfoliatum L.
Perfoliate Penny Cress

Plate 11 RESEDACEAE and CISTACEAE

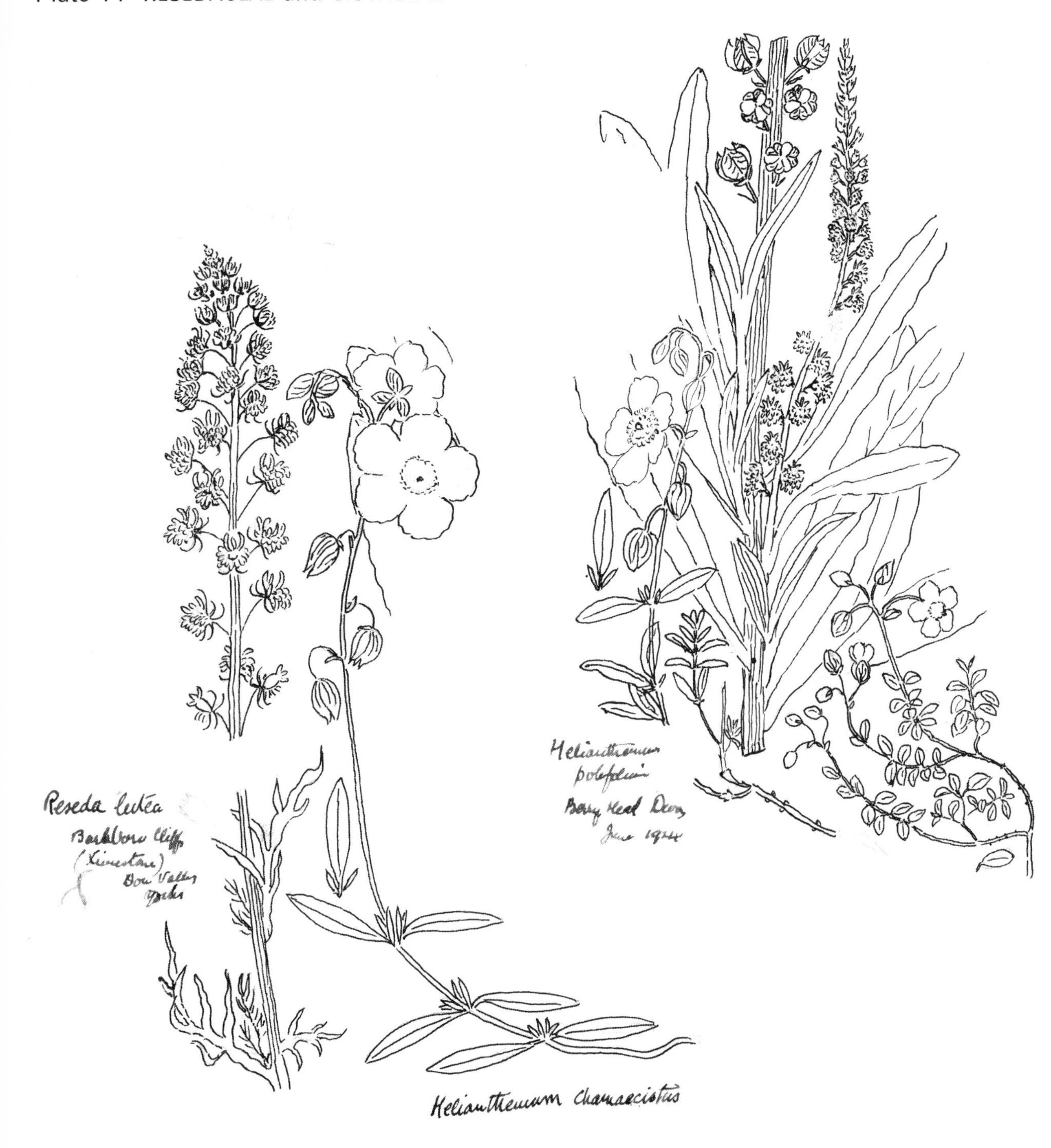

Reseda luteola L.
Dyer's Rocket, Weld

Helianthemum apenninum (L.) Mill.
White Rockrose

Reseda lutea L.
Wild Mignonette

Helianthemum nummularium (L.) Mill.
Common Rockrose

Helianthemum canum (L.) Baumg.
Hoary Rockrose

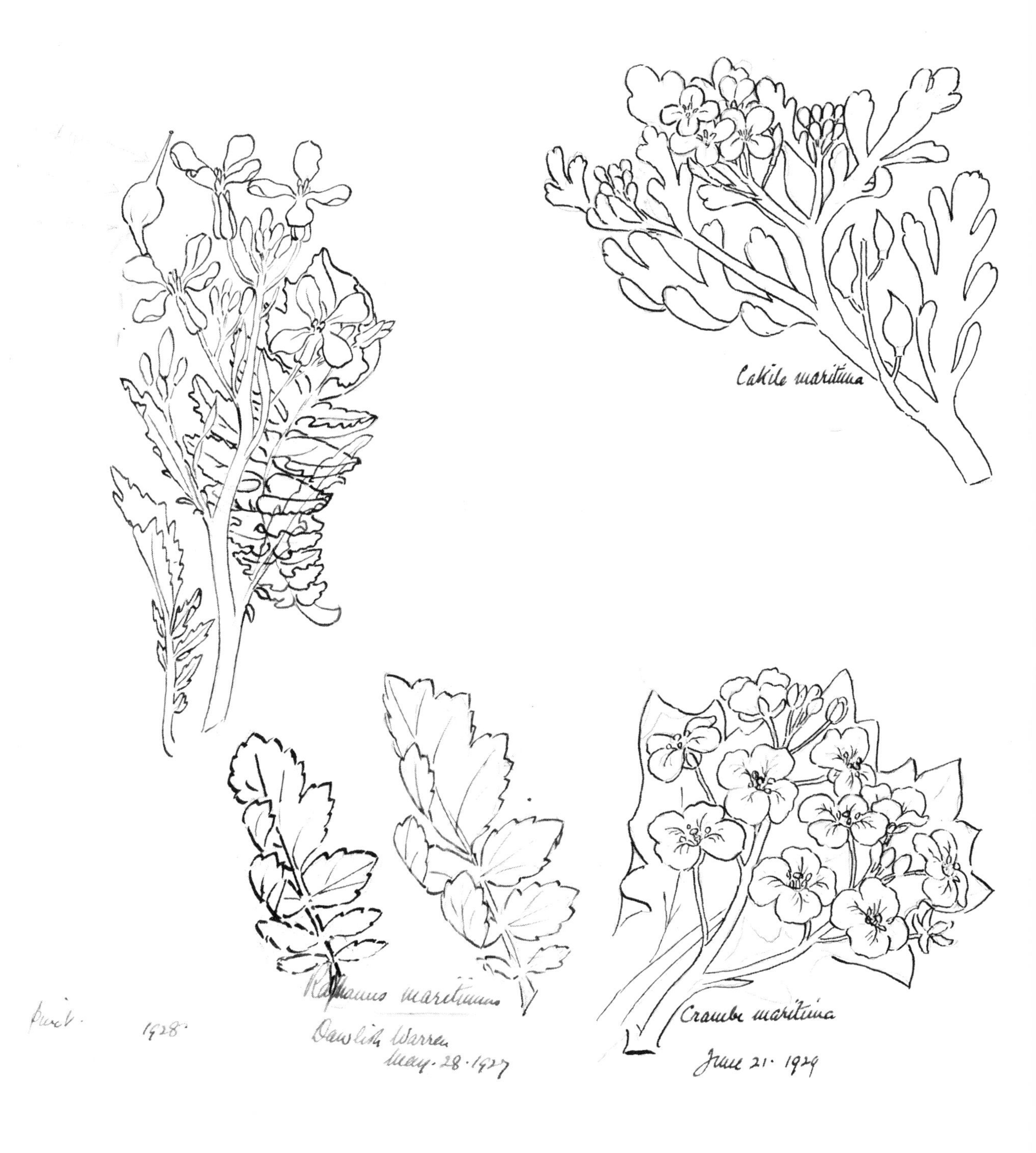

Raphanus maritimus (Sm.) Thell.
Sea Radish

Cakile maritima Scop.
Sea Rocket

Crambe maritima L.
Sea Kale

Plate 12 VIOLACEAE

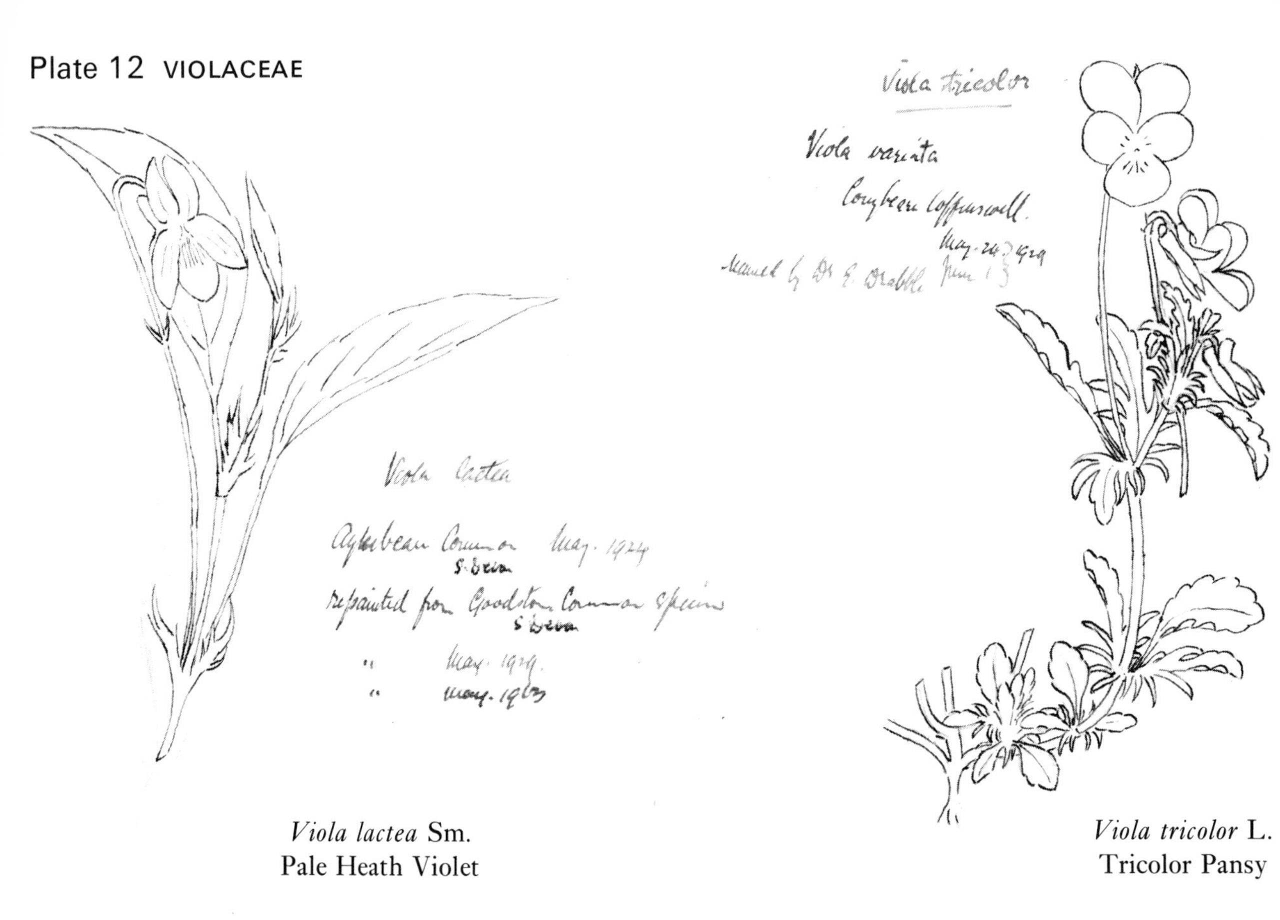

Viola lactea Sm.
Pale Heath Violet

Viola tricolor L.
Tricolor Pansy

Redrawn from original sketches for the *Flora*

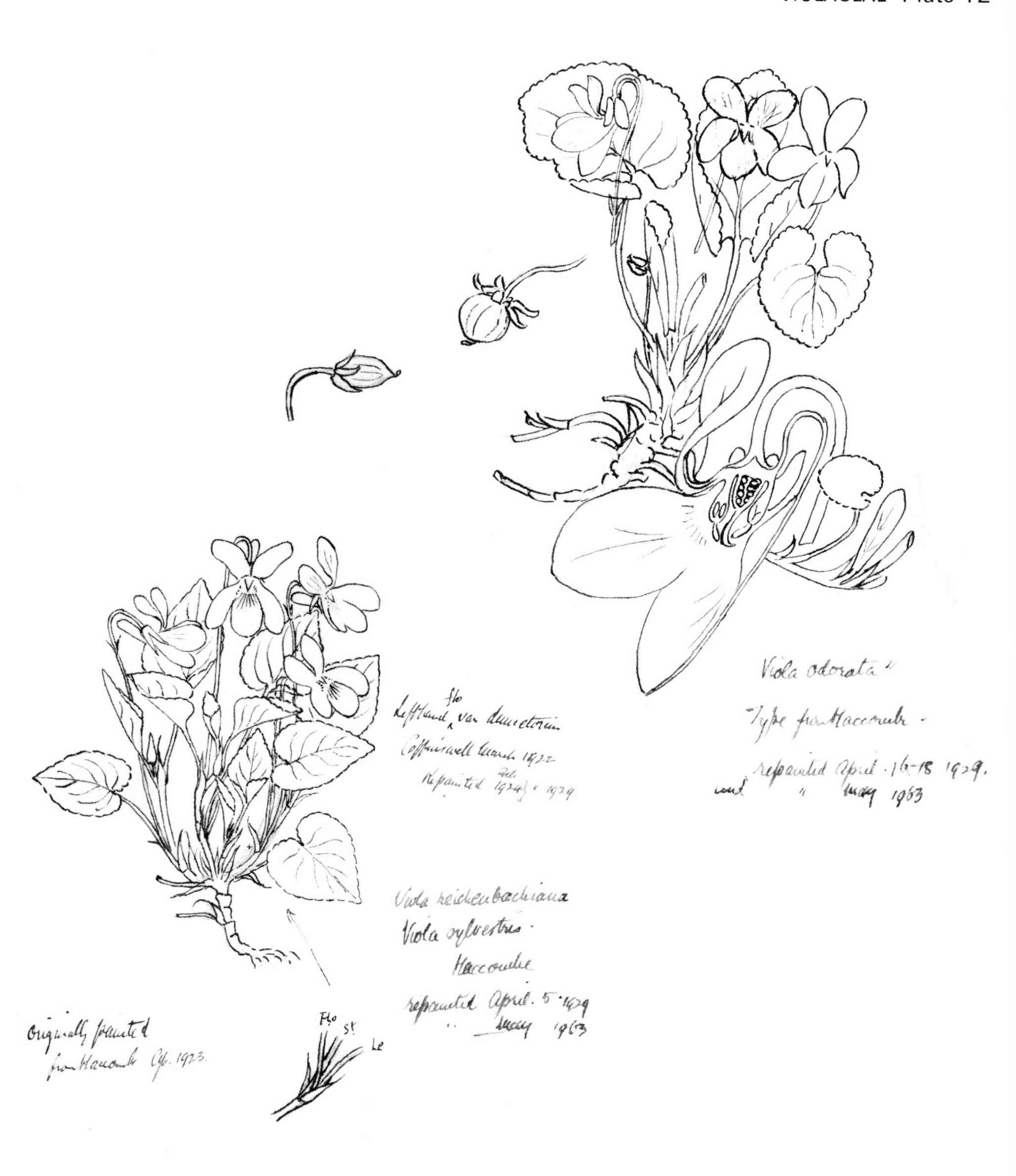

Viola reichenbachiana Jord. ex Bor.
Woodland Violet

Viola odorata L.
Sweet Violet

Silene nutans L.
Nodding or Nottingham Catchfly

Saponaria officinalis L.
Soapwort

Vaccaria pyramidata Medic.

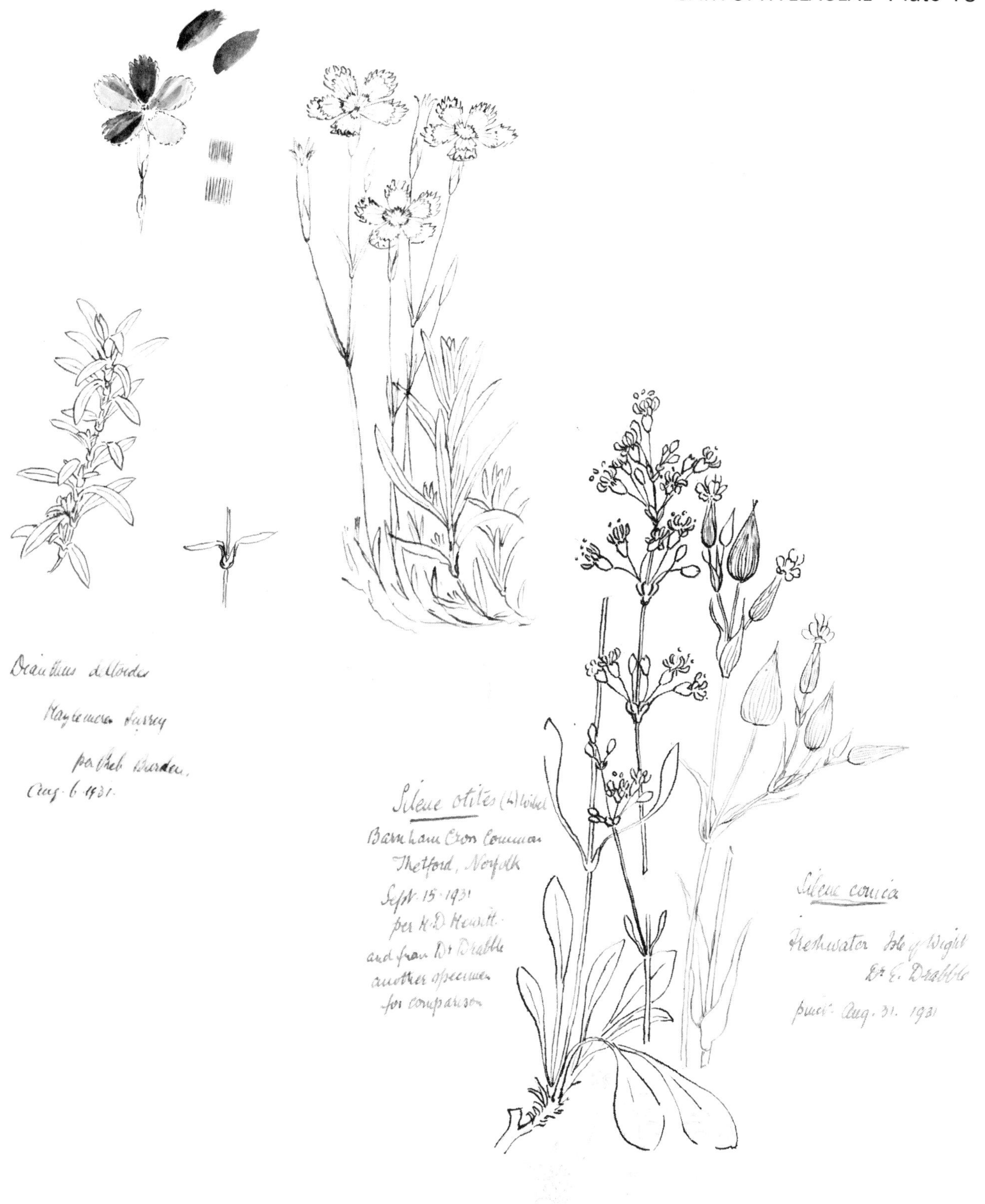

Dianthus deltoides L.
Maiden Pink

Silene otites (L.) Wibel.
Spanish Catchfly

Silene conica L.
Striated Catchfly

Lychnis flos-cuculi L.
Ragged Robin

Silene alba (Mill.) E. H. L. Krause.
White Campion

Silene dioica (L.) Clairv.
Red Campion

Silene noctiflora L.
Night-flowering Catchfly

Separated Sept. 1929.

Moehringia trinervia (L.) Clairv.
Three-veined Sandwort

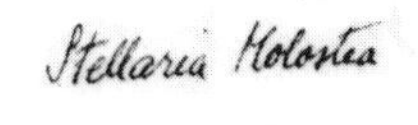

Coffinswell April 28 1924.
Redrawn & Adapted to Exwick Specimens
April 25th 1983
[illegible]

Stellaria holostea L.
Greater Stitchwort

Arenaria gothica Ribblehead Ingleboro.
[illegible] Natural Herbarium

Arenaria norvegica

Arenaria norvegica Gunn.
Arctic Sandwort

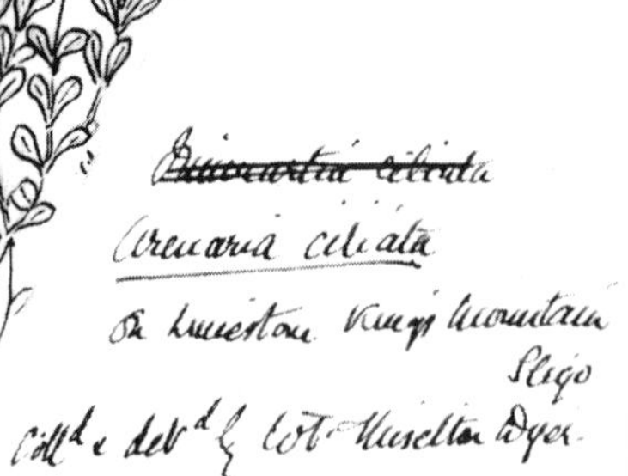

Arenaria ciliata L.
Irish Sandwort

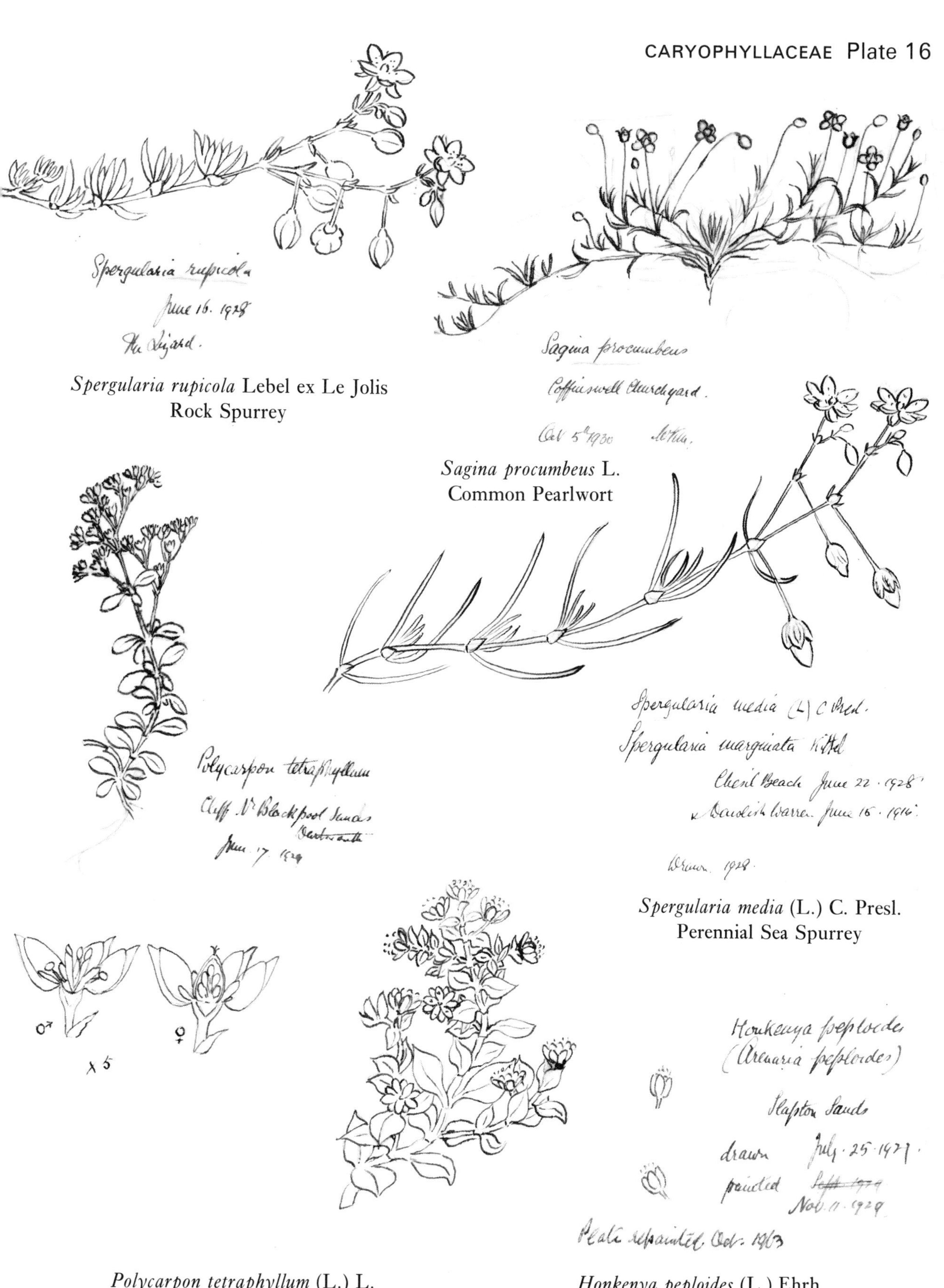

Spergularia rupicola Lebel ex Le Jolis
Rock Spurrey

Sagina procumbeus L.
Common Pearlwort

Spergularia media (L.) C. Presl.
Perennial Sea Spurrey

Polycarpon tetraphyllum (L.) L.
Four-leaved Allseed

Honkenya peploides (L.) Ehrh
Sea Purslane

Plate 17 HYPERICACEAE

Hypericum undulatum Schousb. ex Willd.
Wavy St John's Wort

Hypericum perforatum L.
Perforate St John's Wort

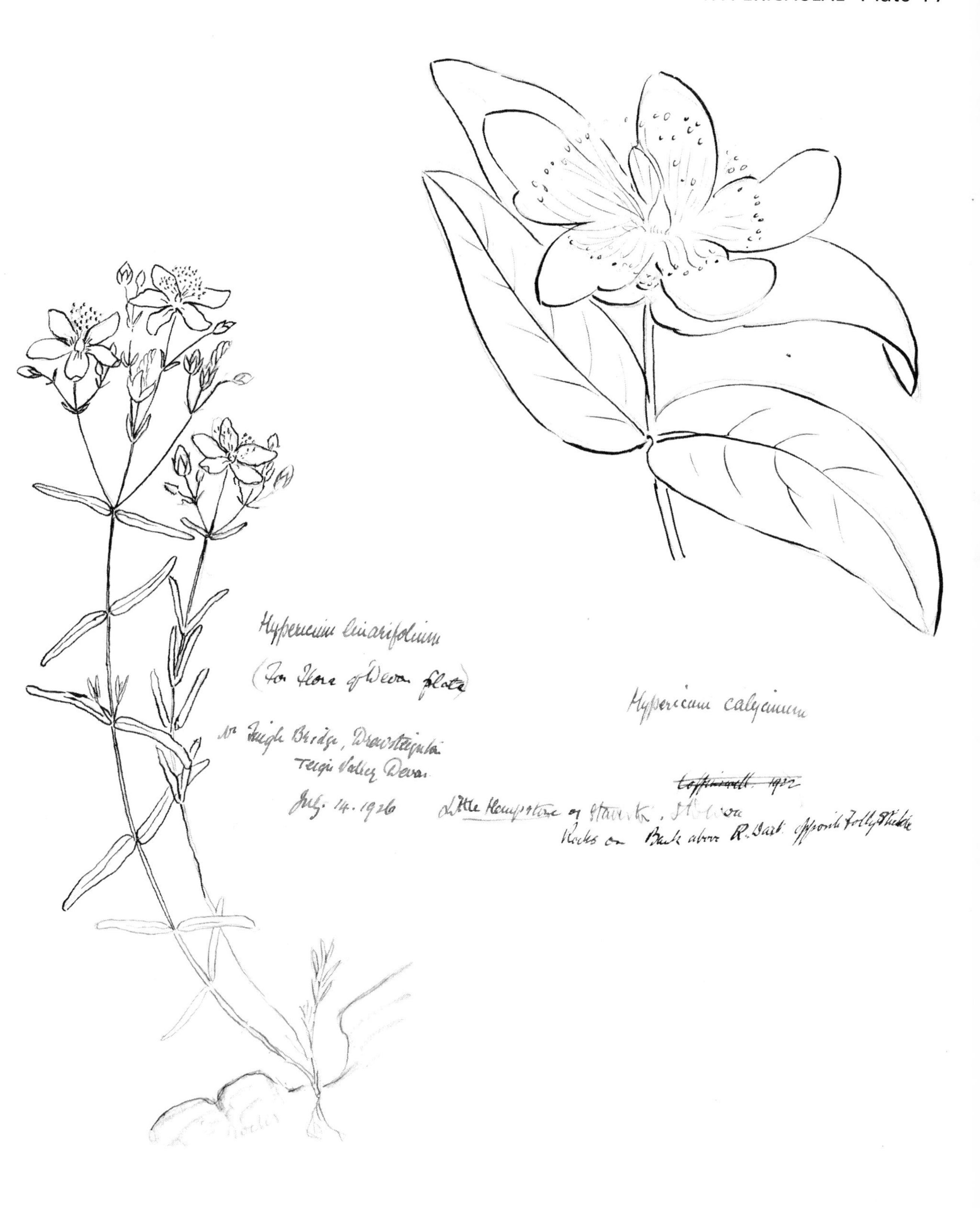

Hypericum linarifolium Vahl.
Narrow-leaved St John's Wort

Hypericum calycinum L.
Rose of Sharon

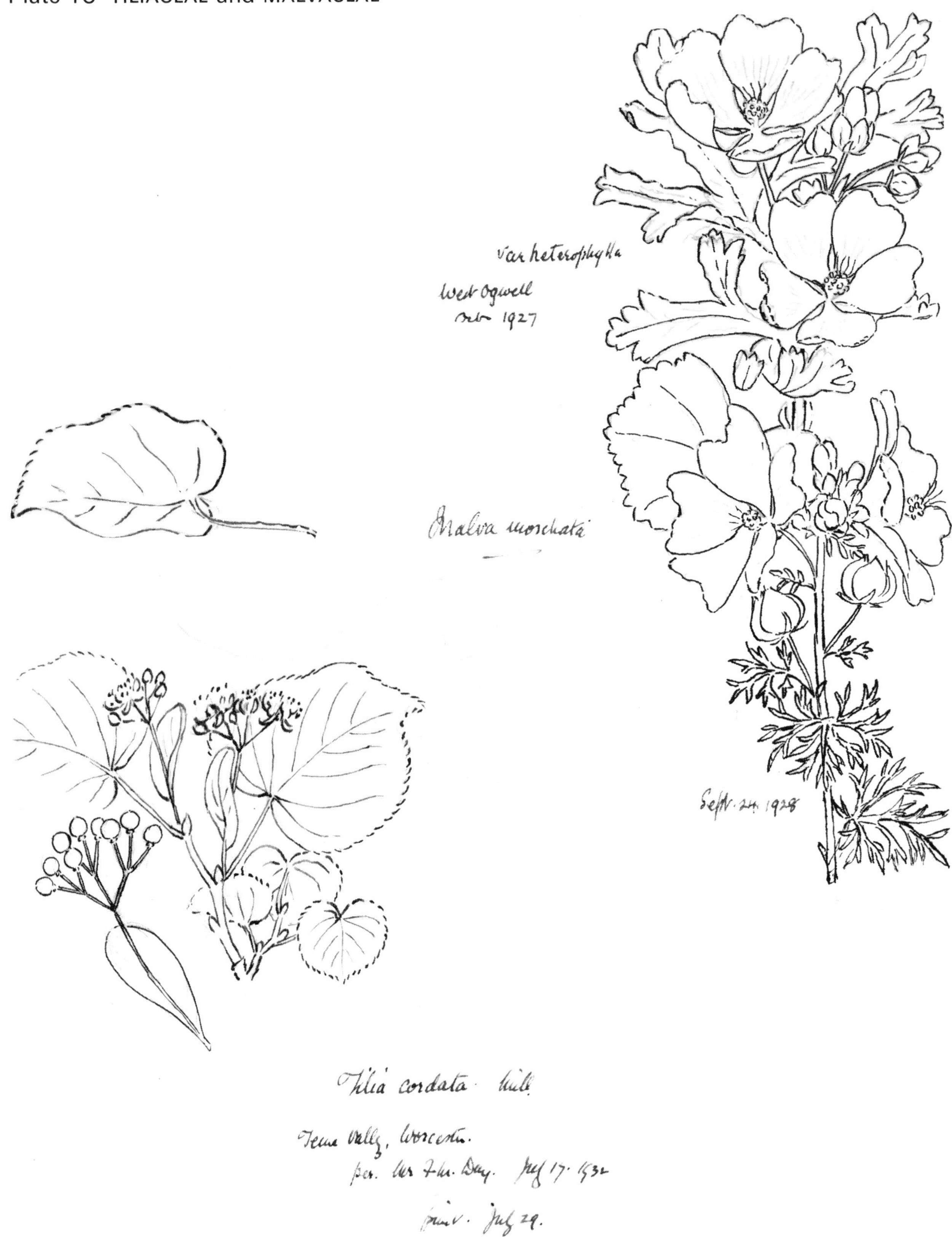

Tilia cordata Mill.
Small-leaved Lime

Malva moschata L.
Musk Mallow

Althaea officinalis L.
Marsh Mallow

Althaea hirsuta L.
Hispid Marsh Mallow

Lavatera arborea L.
Tree Mallow

Geranium pusillum
Dartington Near Top Gate (Broadlears)
Sept. 1925.
repainted Sept. 12. 1932

Geranium pusillum L.
Small-flowered Cranesbill

Geranium sanguineum L.
Blood-red Geranium

Geranium sylvaticum L.
Wood Cranesbill

Linum perenne Mill.
Blue Flax

Plate 20 BALSAMINACEAE and GERANIACEAE

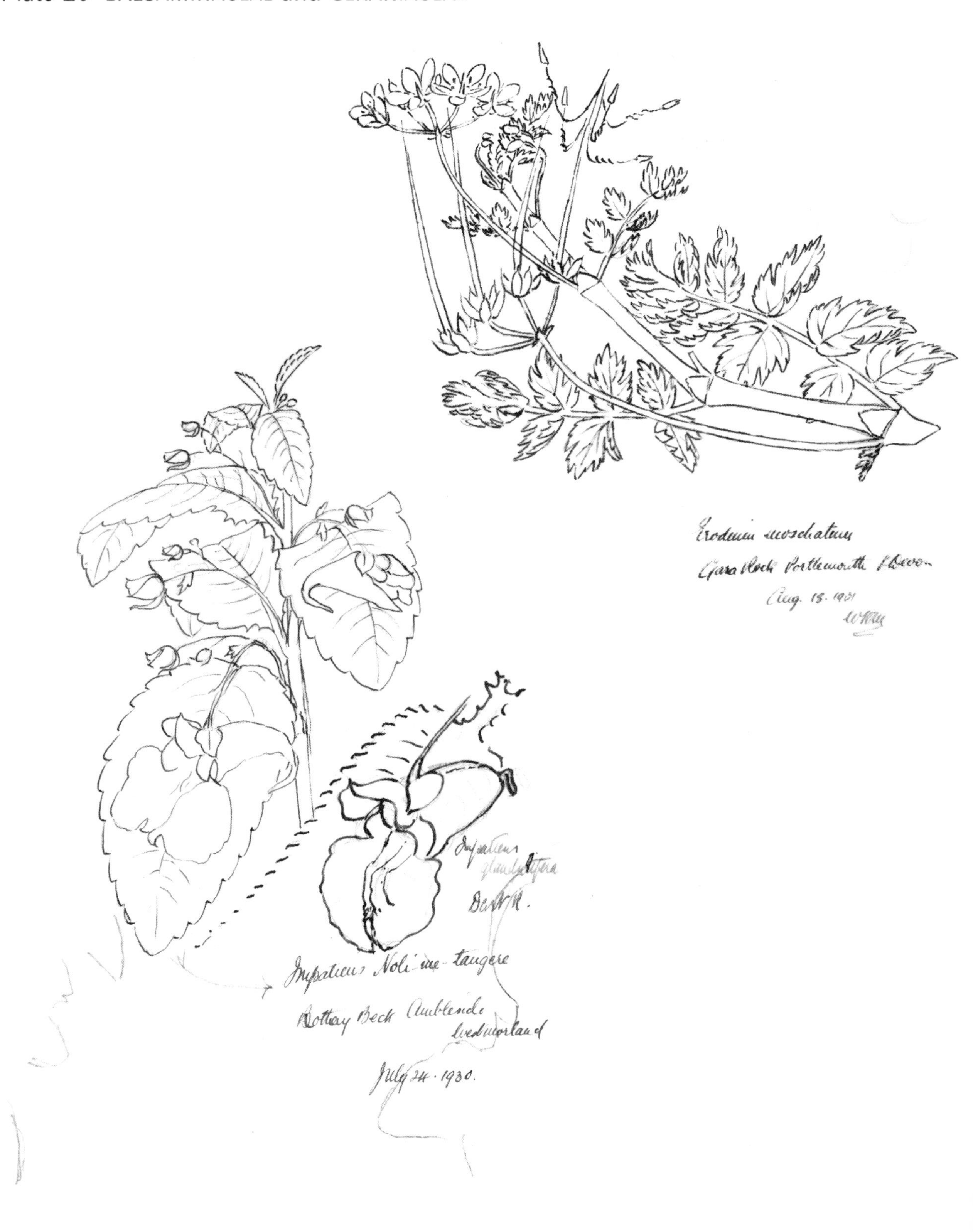

Impatiens noli-tangere L.
Wild Balsam, Touch-me-not

Erodium moschatum (L.) L'Hérit.
Musky Storksbill

Euonymus europaeus

(Ottery 1916.)
Coffinswell June & Oct. 1926

Erodium cicutarium L'Herit.
form triviale

Milber Down Coffinswell
above Aller Sand pit.
August 29th 1932

Erodium cicutarium (L.) L'Hérit.
Common Storksbill

Euonymus europaeus L.
Spindle

Plate 21 PAPILIONACEAE

Ononis repens L. Var. *horrida* Lange
Restharrow

Medicago sativa L.
Lucerne

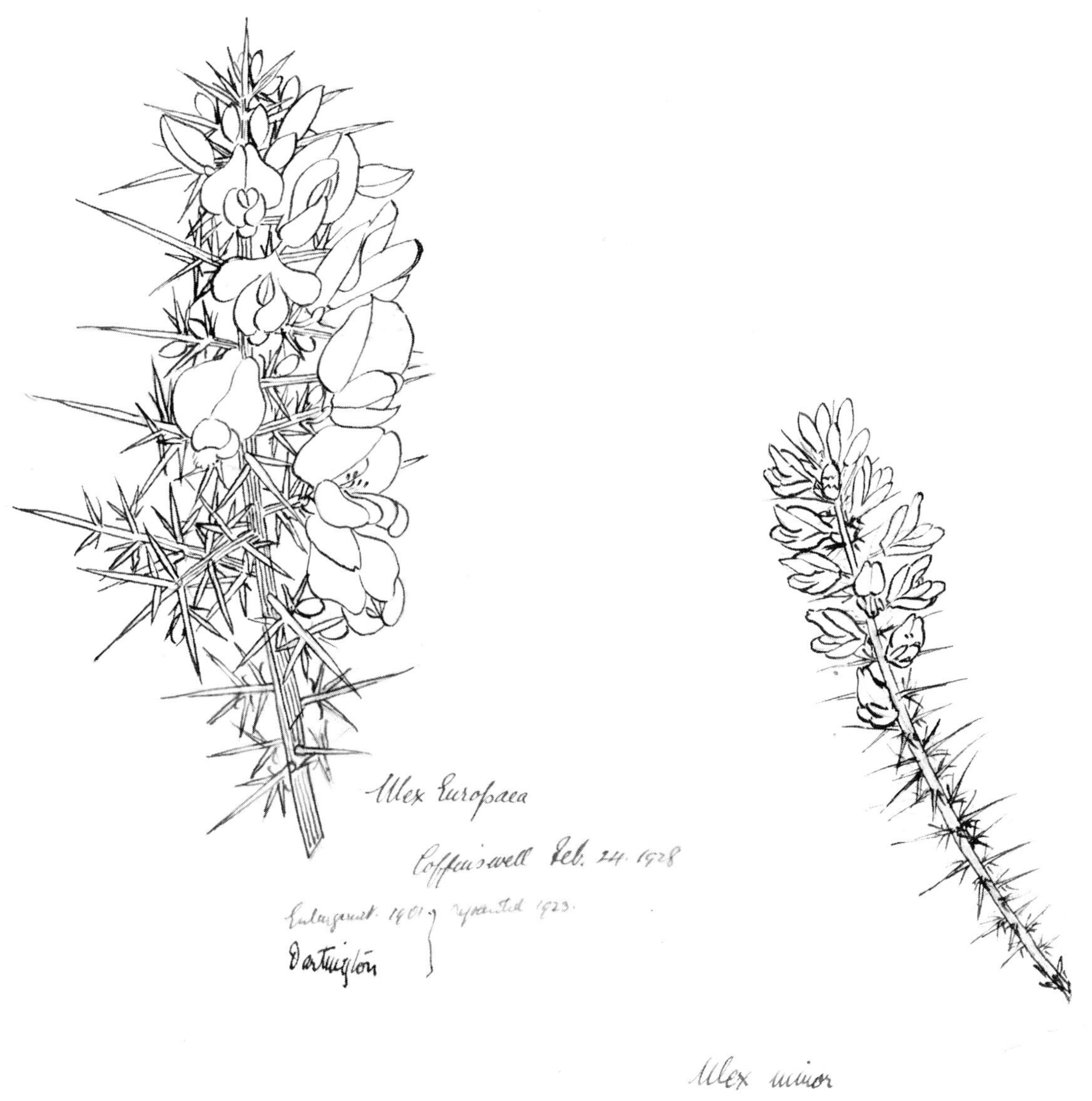

Ulex europaeus L.
Common Gorse, Furze

Ulex minor Roth.
Small Furze

Plate 22 PAPILIONACEAE

Meliotus altissima Thuill.
Yellow Melilot

Trifolium molinerii Balb. ex Hornem
Large Lizard Clover

Anthyllis vulneraria L.
Kidney Vetch

Lotus uliginosus Schkuhr.
Marsh Bird's-foot Trefoil

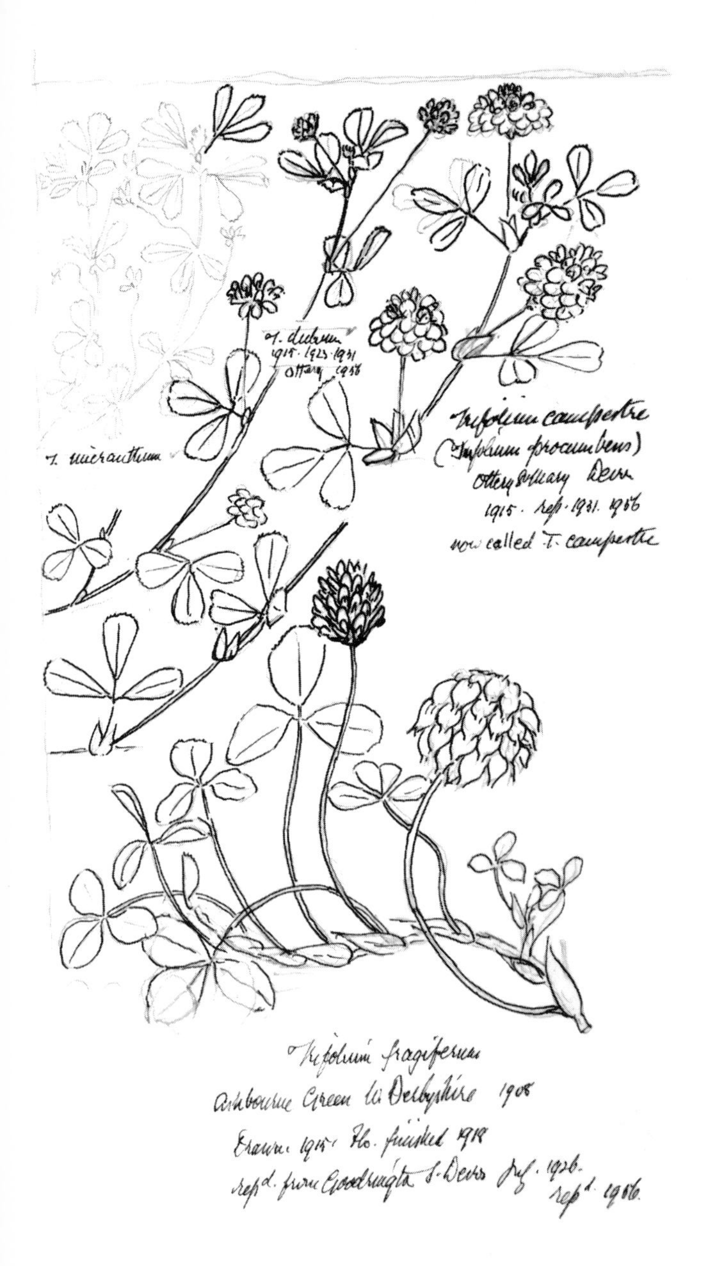

Trifolium campestre Schreb.
Hop Trefoil

Trifolium fragiferum L.
Strawberry-headed clover

Trifolium micanthum Viv.
Least Yellow Trefoil

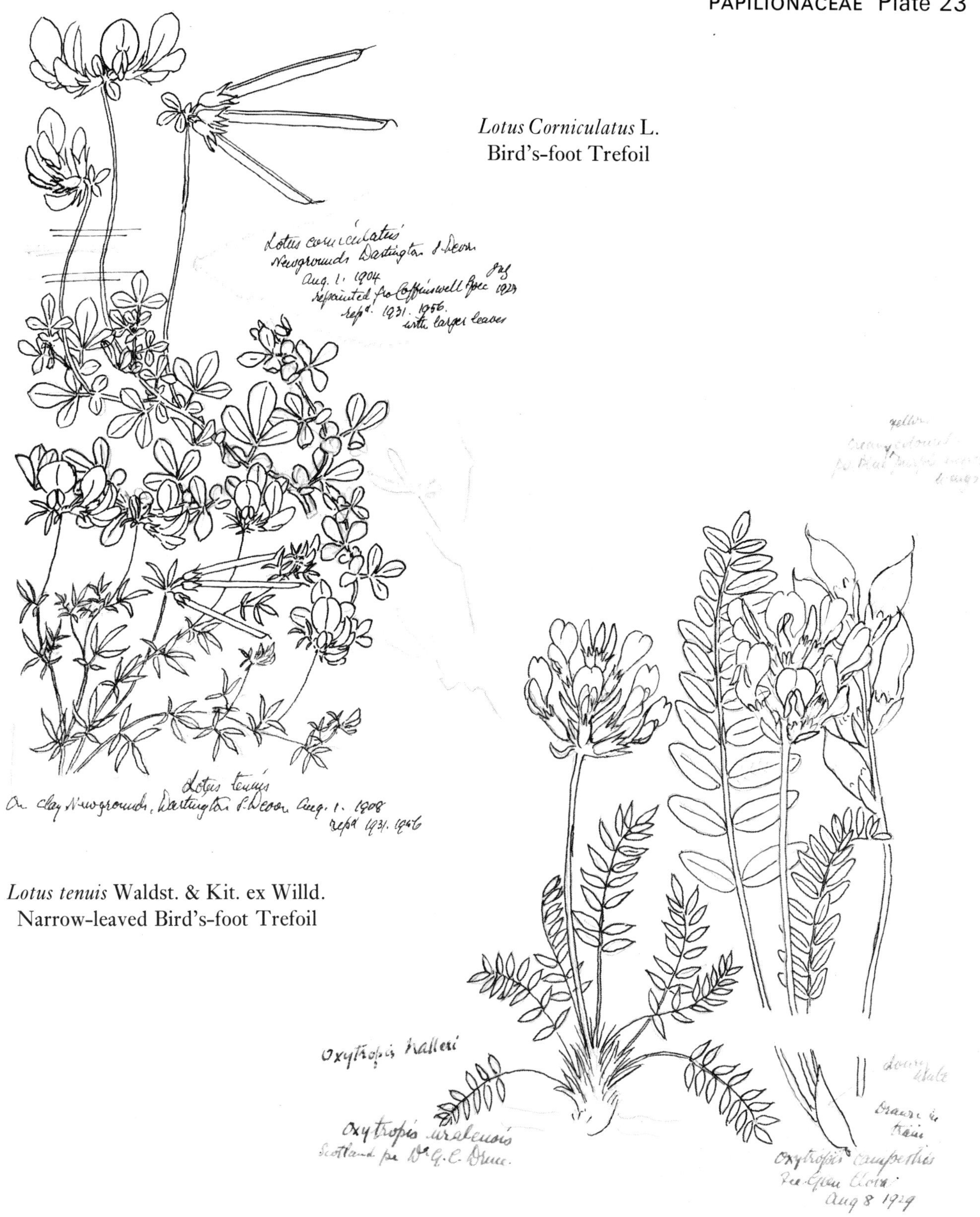

Lotus Corniculatus L.
Bird's-foot Trefoil

Lotus tenuis Waldst. & Kit. ex Willd.
Narrow-leaved Bird's-foot Trefoil

Oxytropis halleri Bunge.

Oxytropis campestris (L.) DC.

Plate 24 PAPILIONACEAE

Onobrychis viciifolia Scop.	*Astragalus glycyphyllos* L.	*Astragalus alpinus* L.
Sainfoin	Wild Liquorice, Milk-vetch	Alpine Milk-vetch

Vicia hirsuta Gray

Woodham La. Addlestone Surrey
July 20. 1932

Vicia Orobus

from Cheddar.
per H. S. Thompson
Gathered June. 11. in bud.
Flowered. June 14 - 18'
painted June 17 - 18.
1927

Vicia hirsuta (L.) Gray.
Hairy Tare

Vicia orobus DC.
Wood Bitter Vetch

The figure was improved from a Prawle specimen 1948

Vicia lutea
The Lizard Cornwall
per Mrs Jennings
July 13 · 1928

Vicia angustifolia
Torcross. July 28. 1927
Adapted to specimen from Haccombe May 21. 1928
painted May 1928

Vicia angustifolia L.
Narrow-leaved Vetch

Vicia lutea L.
Yellow Vetch

Lathyrus aphaca (L.) L.
Yellow Vetchling

Lathyrus sylvestris L.
Wild Pea

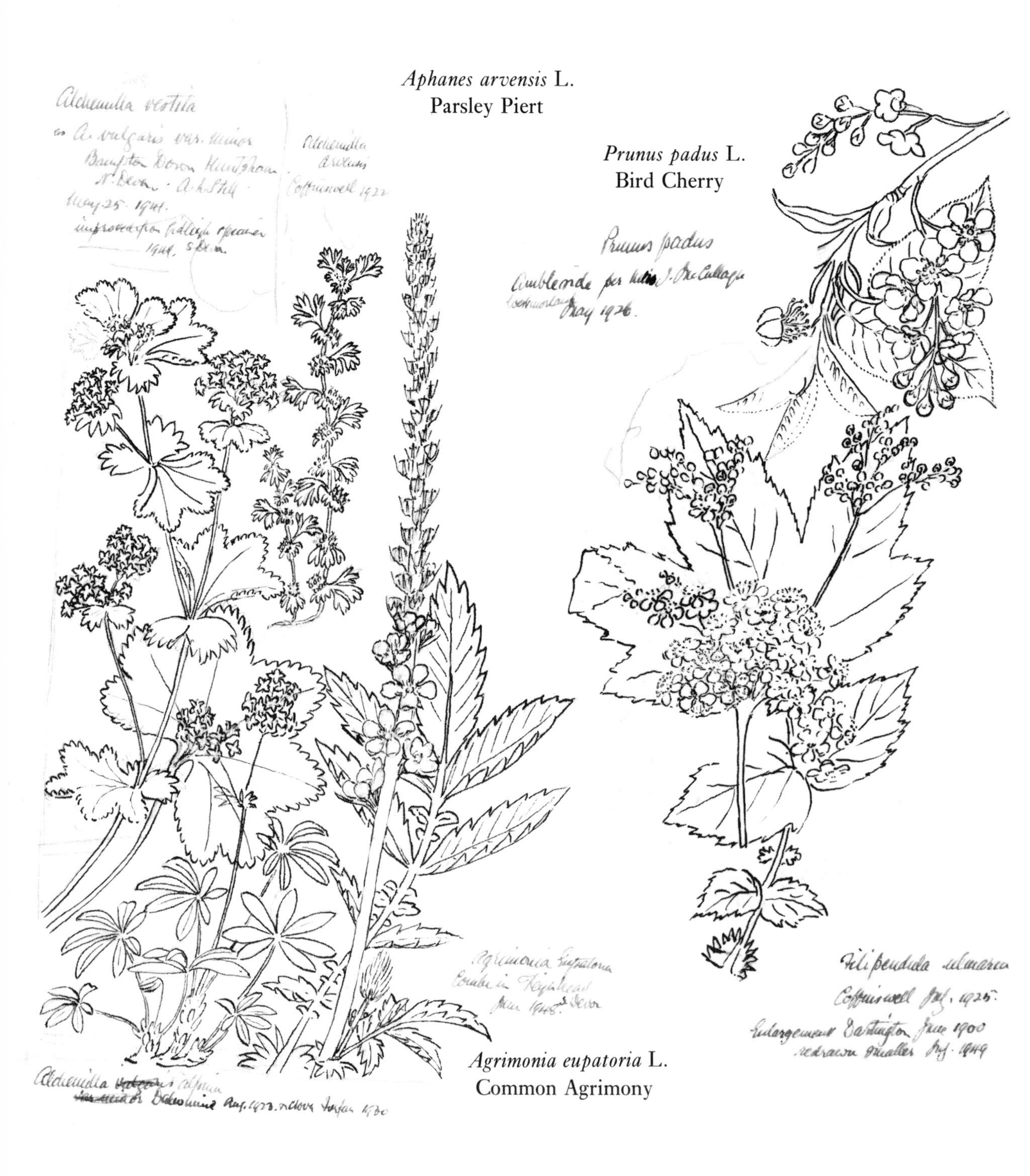

Alchemilla filicaulis Buser. Subsp. *vestita* (Buser) M. E. Bradshaw
Alpine Lady's Mantle

Filipendula ulmaria (L.) Maxim
Meadow Sweet

Prunus spinosa L.
Blackthorn, Sloe
Prunus avium (L.) L.
Gean, Wild Cherry

Prunus insititia (L.) C. K. Schneid.
Bullace

Prunus domestica L.
Wild Plum

Prunus cerasus L.
Sour Cherry

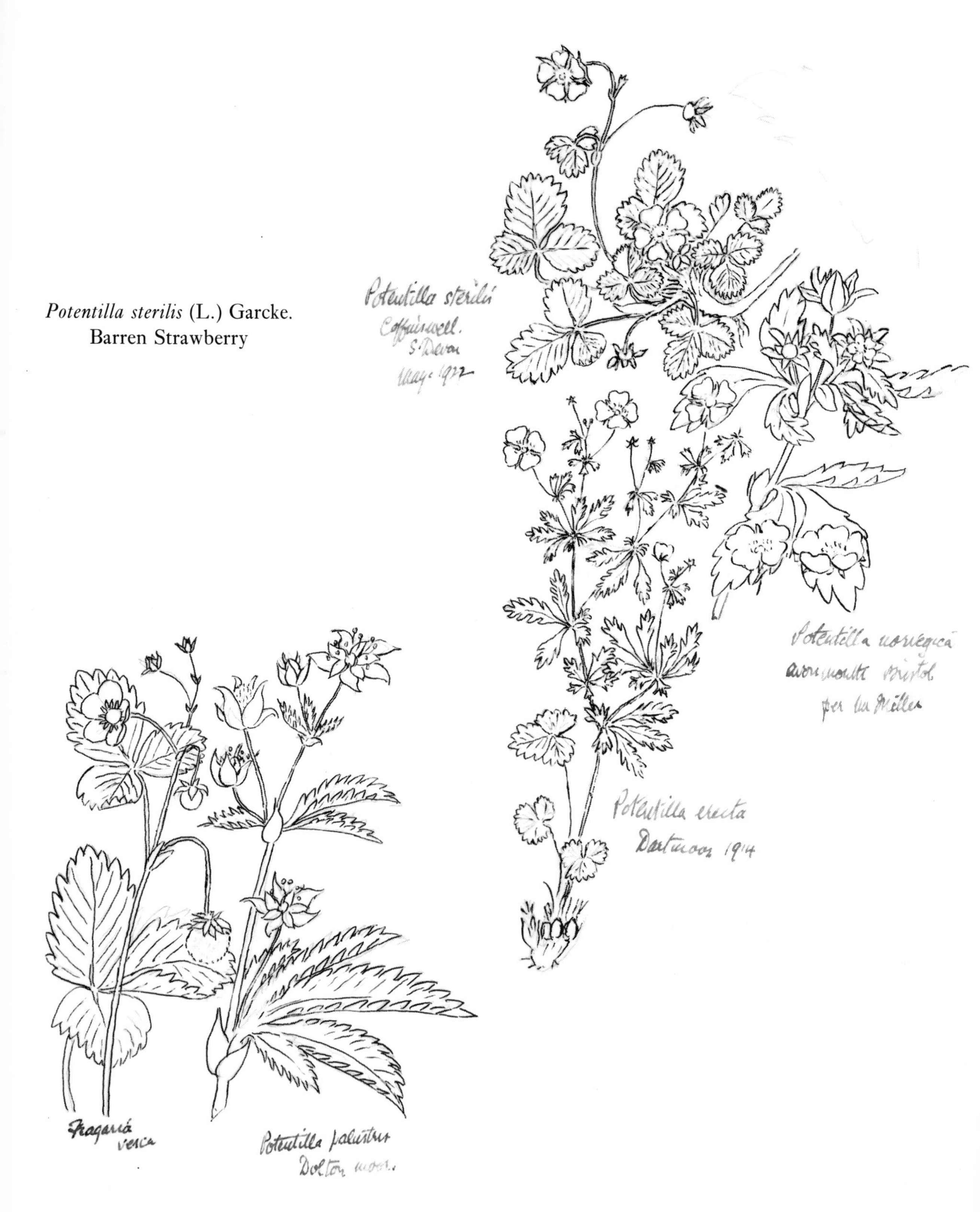

Potentilla sterilis (L.) Garcke.
Barren Strawberry

Fragaria vesca L.
Wild Strawberry

Potentilla palustris (L.) Scop.
Marsh Cinquefoil

Potentilla erecta (L.) Räusch.
Common Tormentil

Potentilla norvegica L.

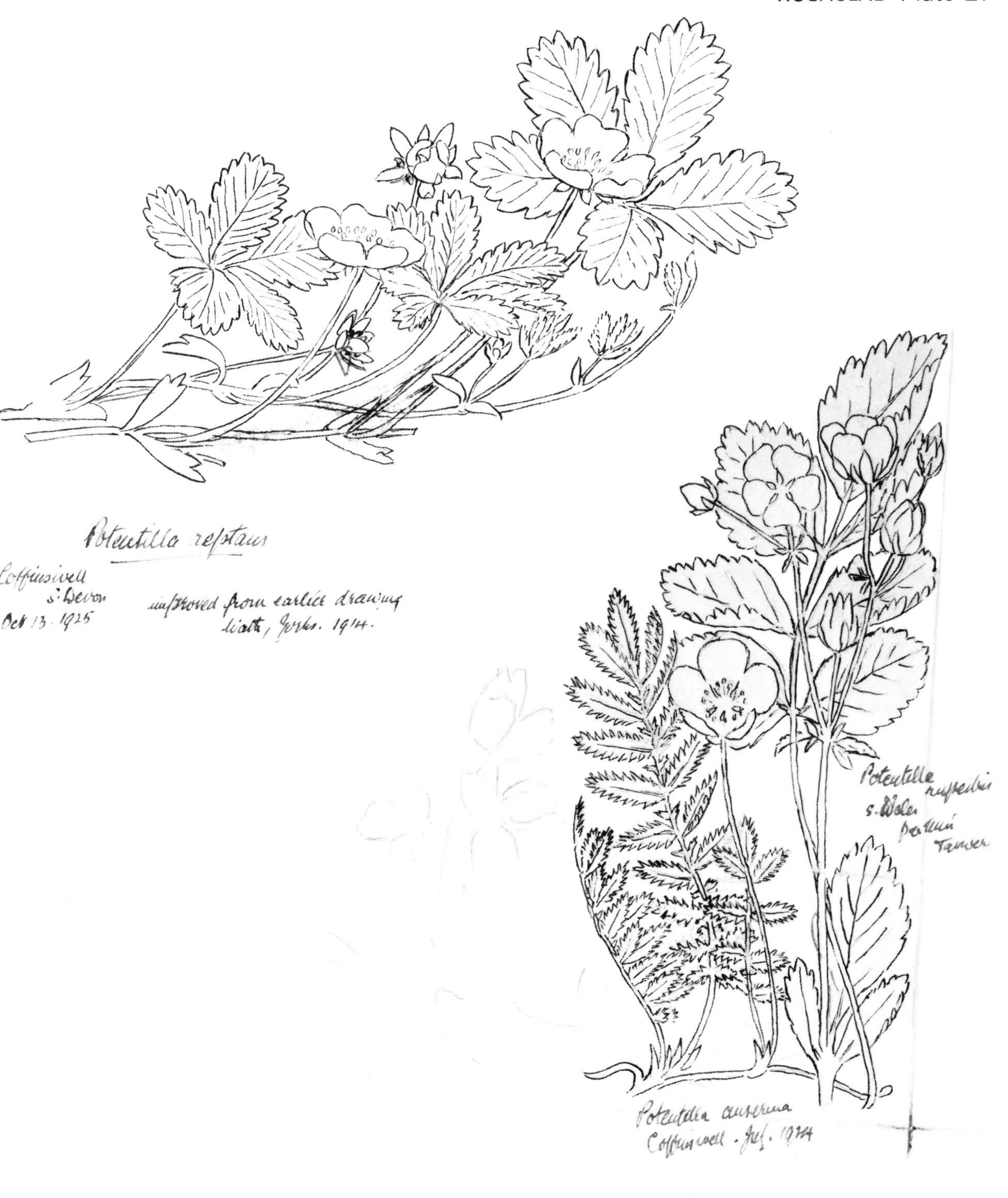

Potentilla reptans L.
Creeping Cinquefoil

Potentilla anserina L.
Silverweed

Potentilla rupestris L.
Rock Cinquefoil

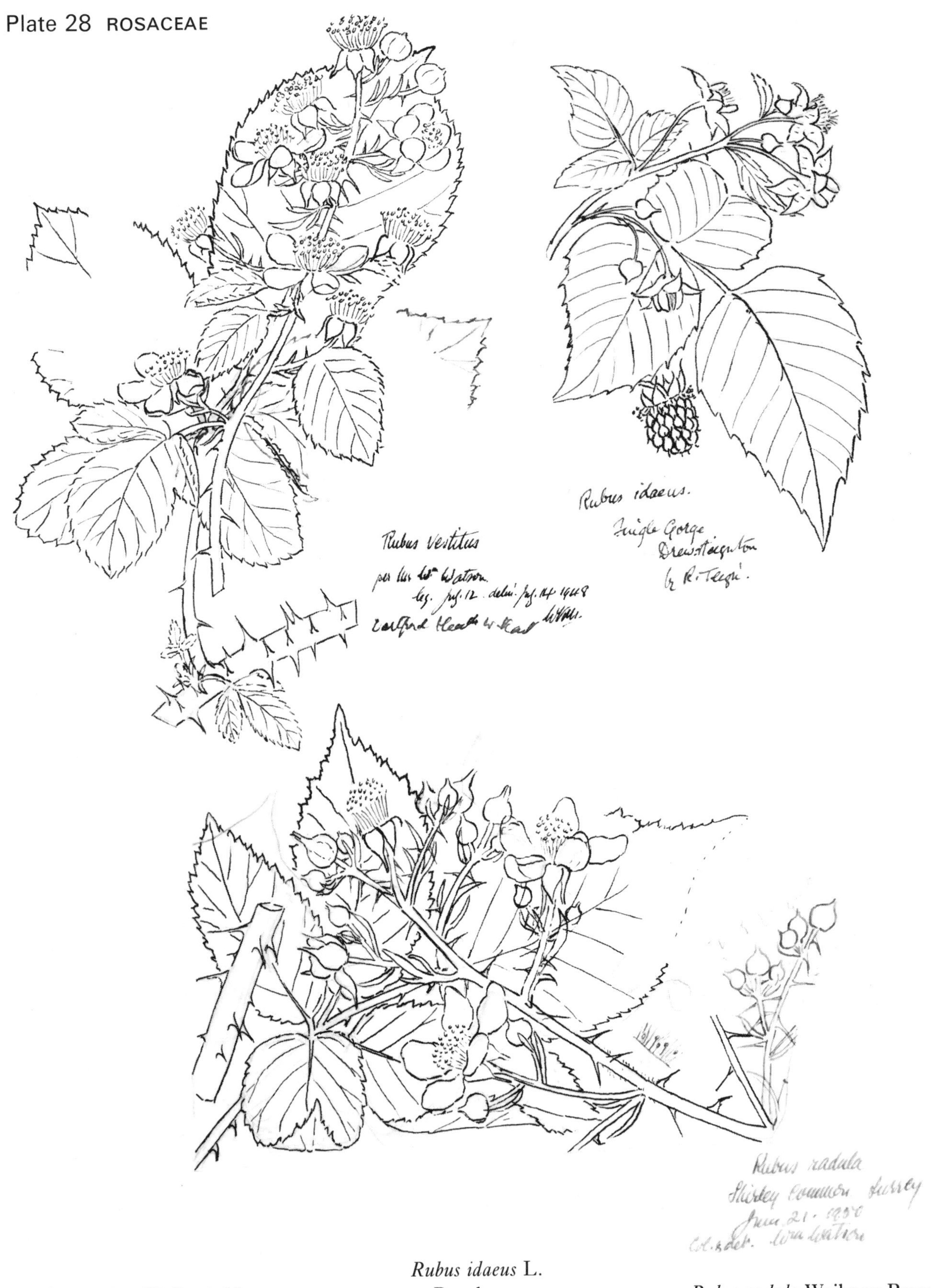

Rubus vestitus Weihe & Nees

Rubus idaeus L.
Raspberry

Rubus radula Weihe ex Boenn.

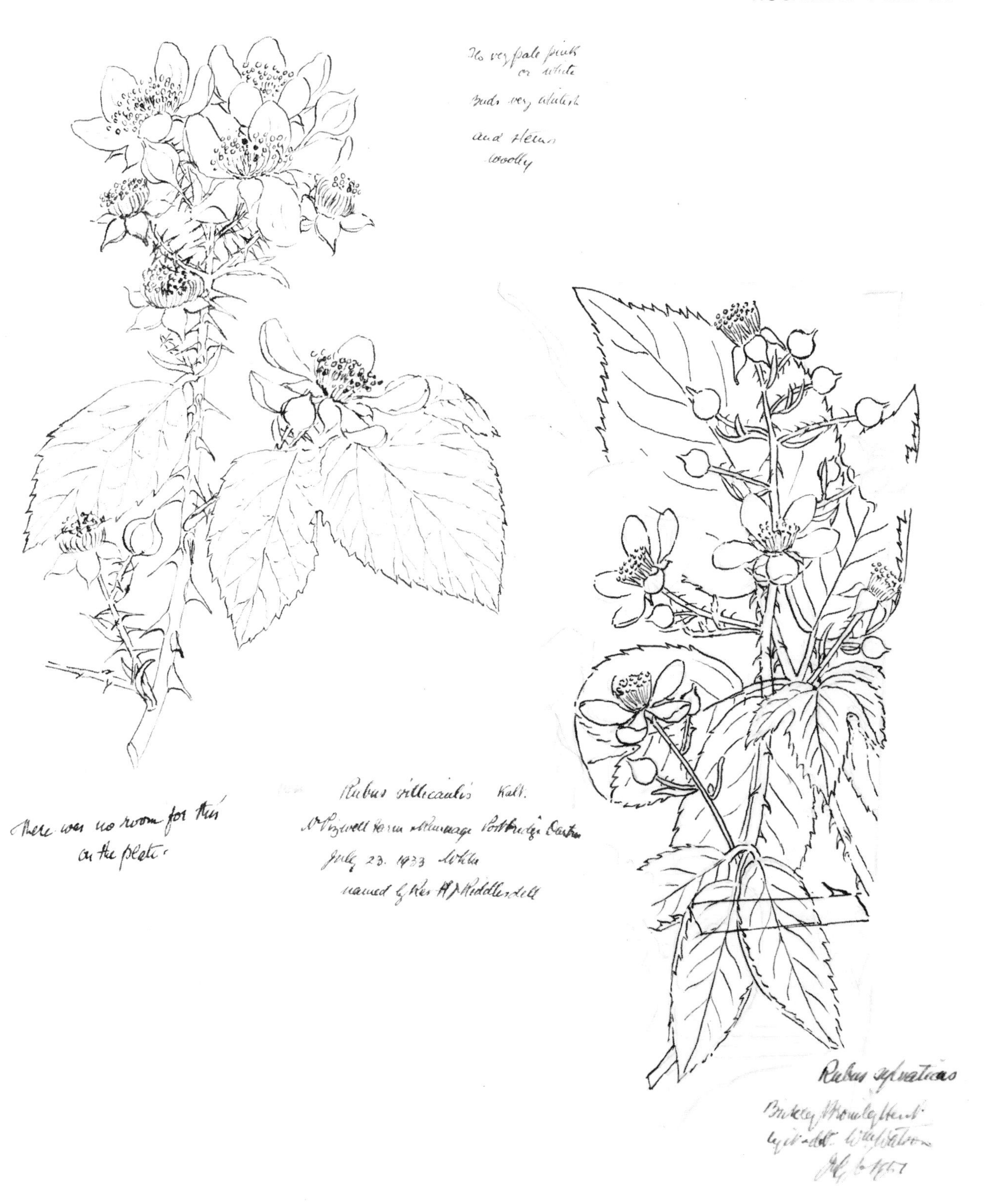

Rubus villicaulis
(not on original plate)

Rubus sylvaticus Weihe & Nees

Rubus bellardii Weihe & Nees

Dryas octopetala L.
Mountain Avens

Painted Aug 6. 1928

a hybrid not pure caesius
Watson

Rubus caesius

Haccombe. Aug. 14 1927.

painted Aug. 15.

1927.

Geum rivale L.
Water Avens

Rubus caesius L.
Dewberry

Rosa canina L.
Dog Rose

Redrawn from original sketches for the *Flora*

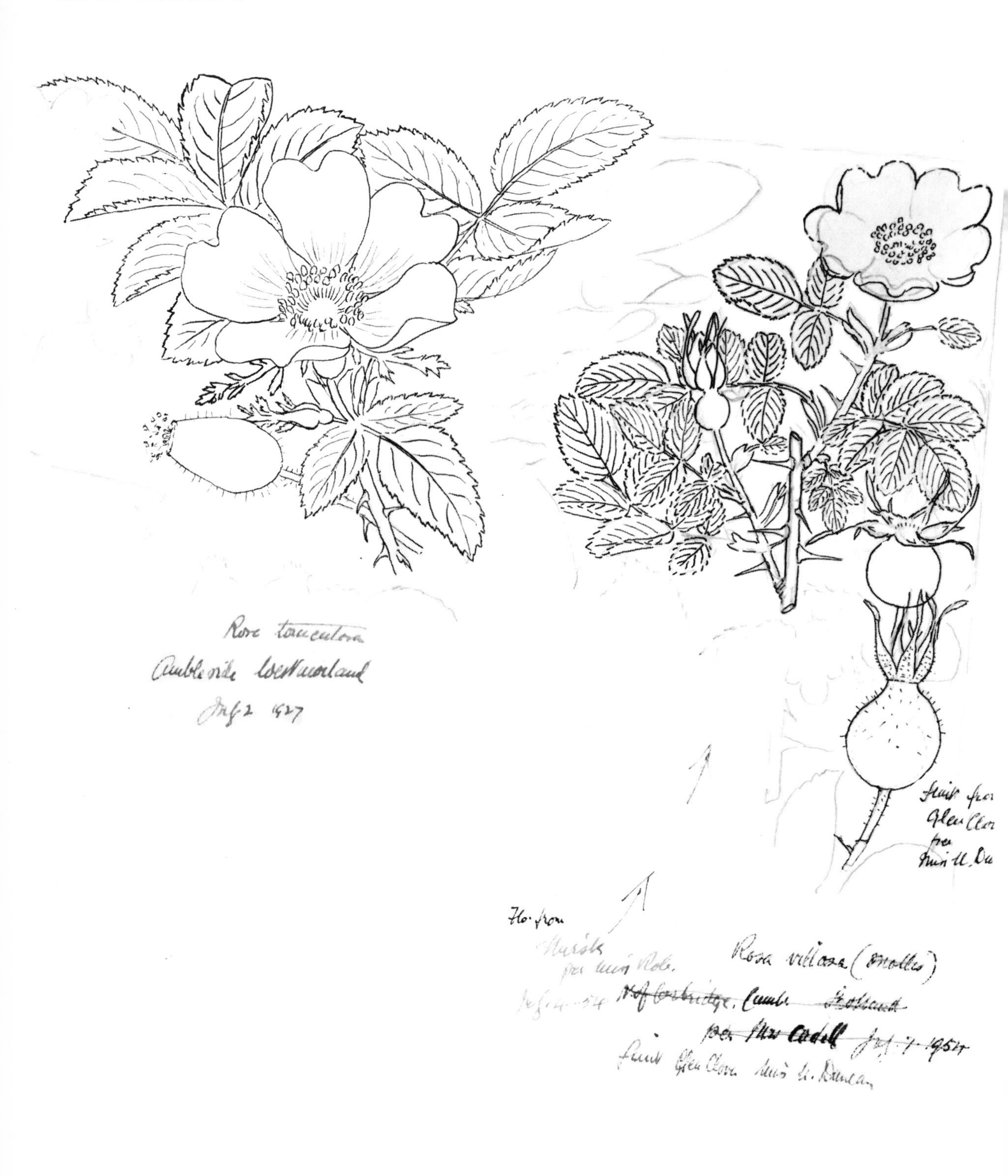

Rosa tomentosa Sm.
Downy Rose

Rosa villosa L.
Soft-leaved Rose

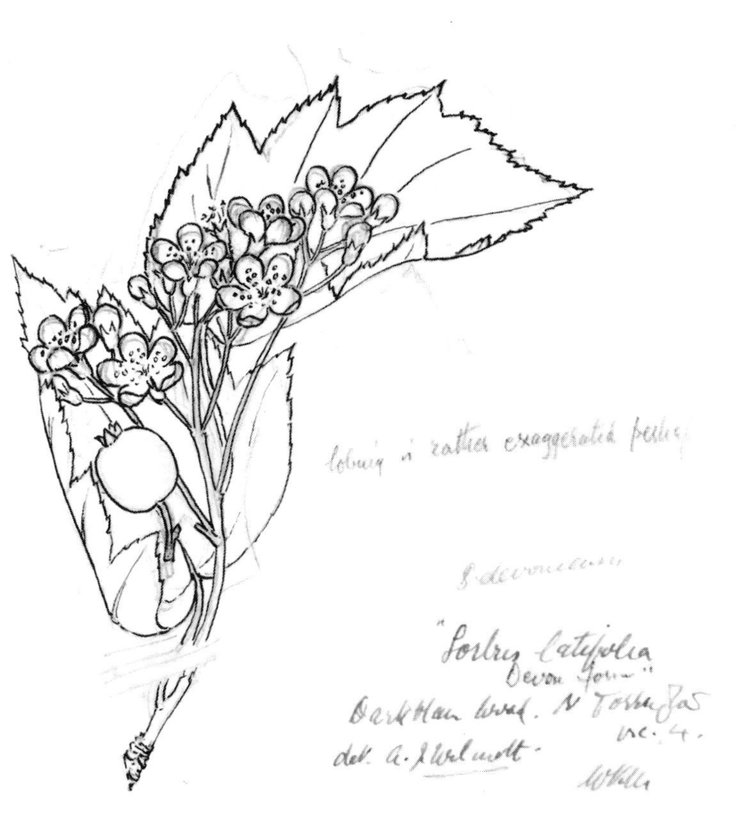

Sorbus latifolia agg.
Broad-leaved White Beam

Sorbus torminalis (L.) Crantz.
Wild Service Tree

Sorbus aria agg.
White Beam

Pyrus pyraster Burgsd.
Pear

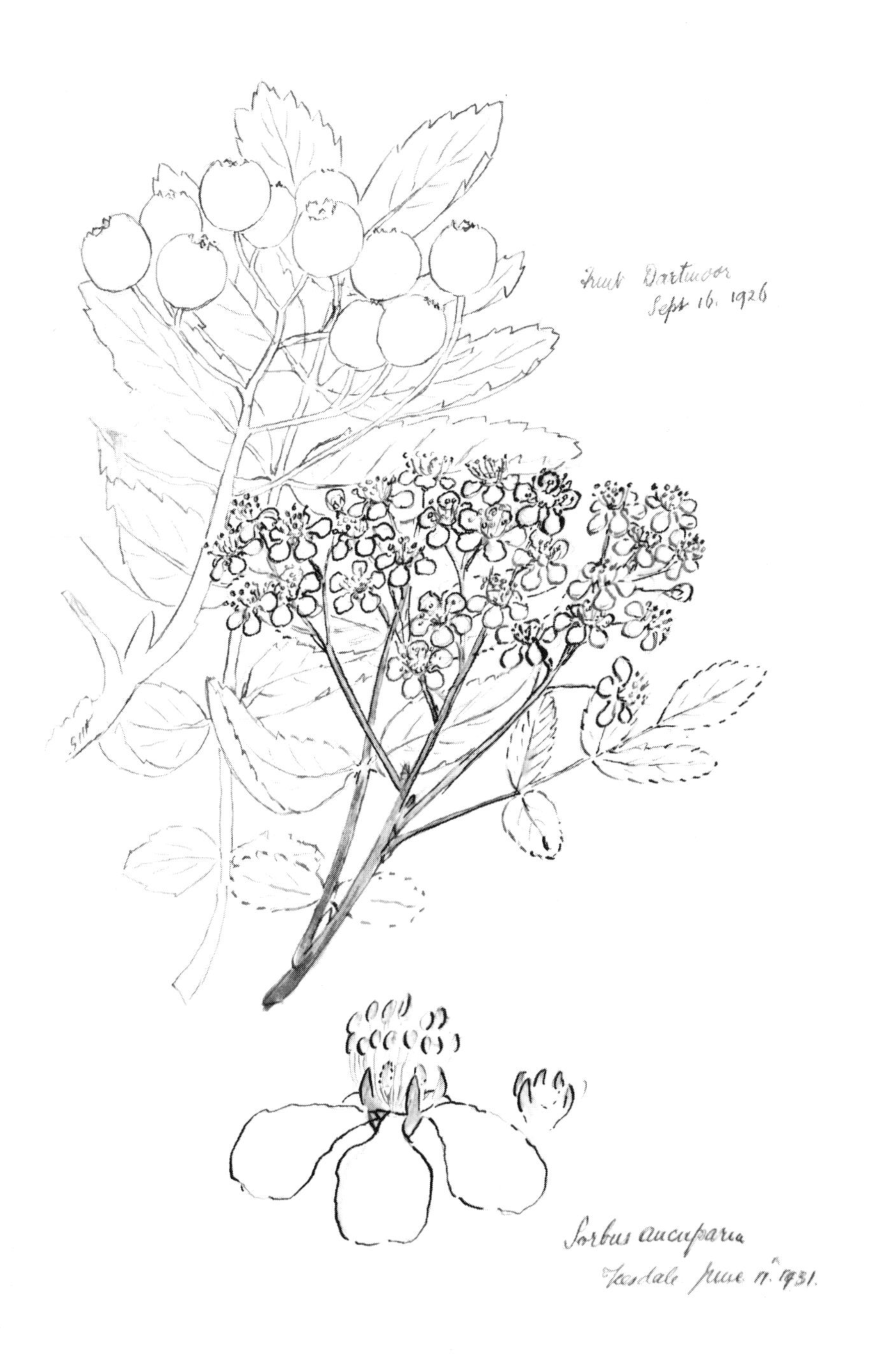

Sorbus aucuparia L.
Rowan, Mountain Ash

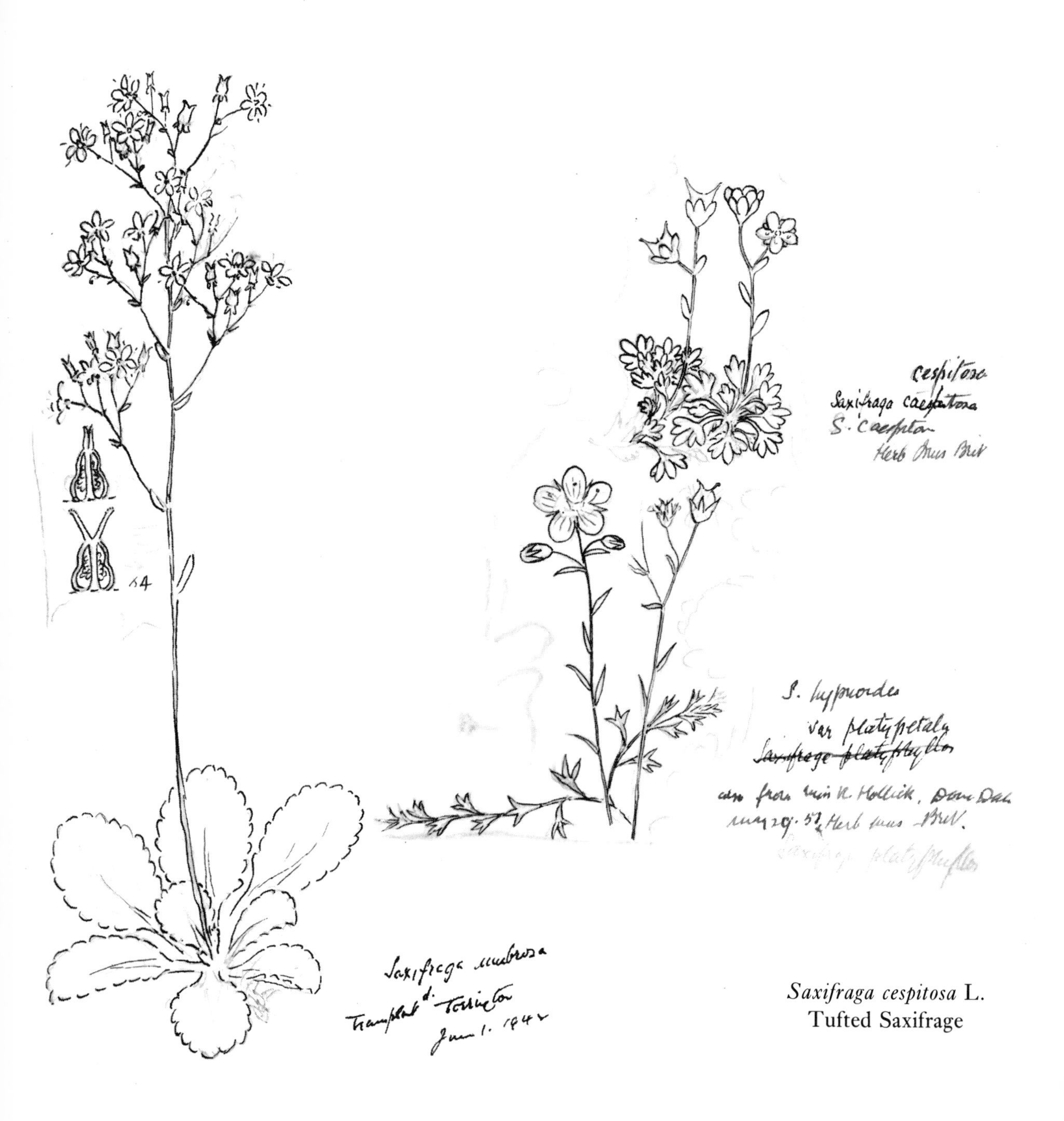

Saxifraga cespitosa L.
Tufted Saxifrage

Saxifraga x urbium D. A. Webb.
London Pride

Saxifraga hypnoides L. Var. *platypetala* Sm.
Mossy Saxifrage.

Saxifraga granulata L.
Meadow Saxifrage, Bulbous Saxifrage

Saxifraga oppositifolia L.
Purple Saxifrage

Plate 33 GROSSULARIACEAE and CRASSULACEAE

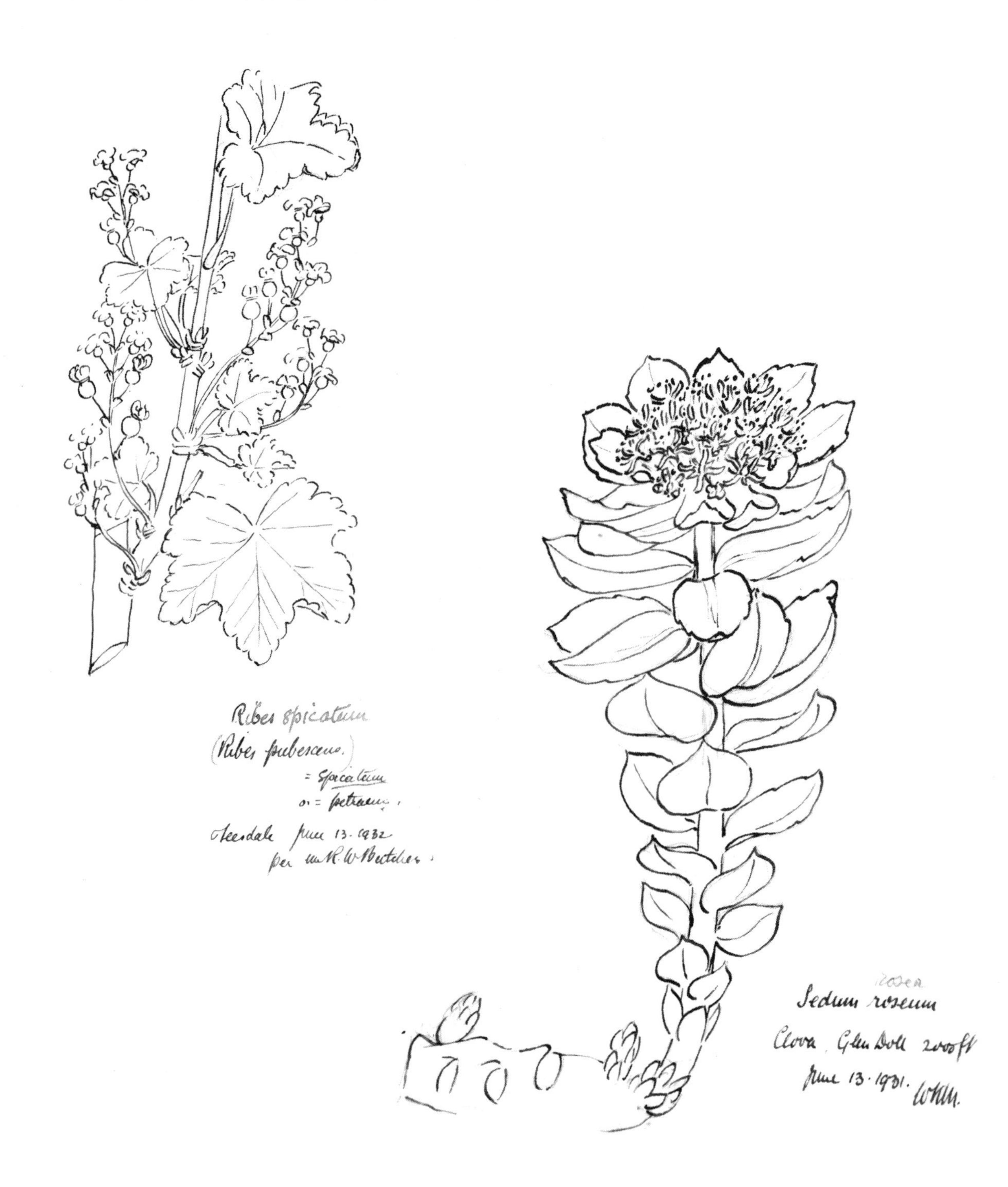

Ribes spicatum Robson.
Upright Red Currant

Rhodiola rosea L.
Roseroot

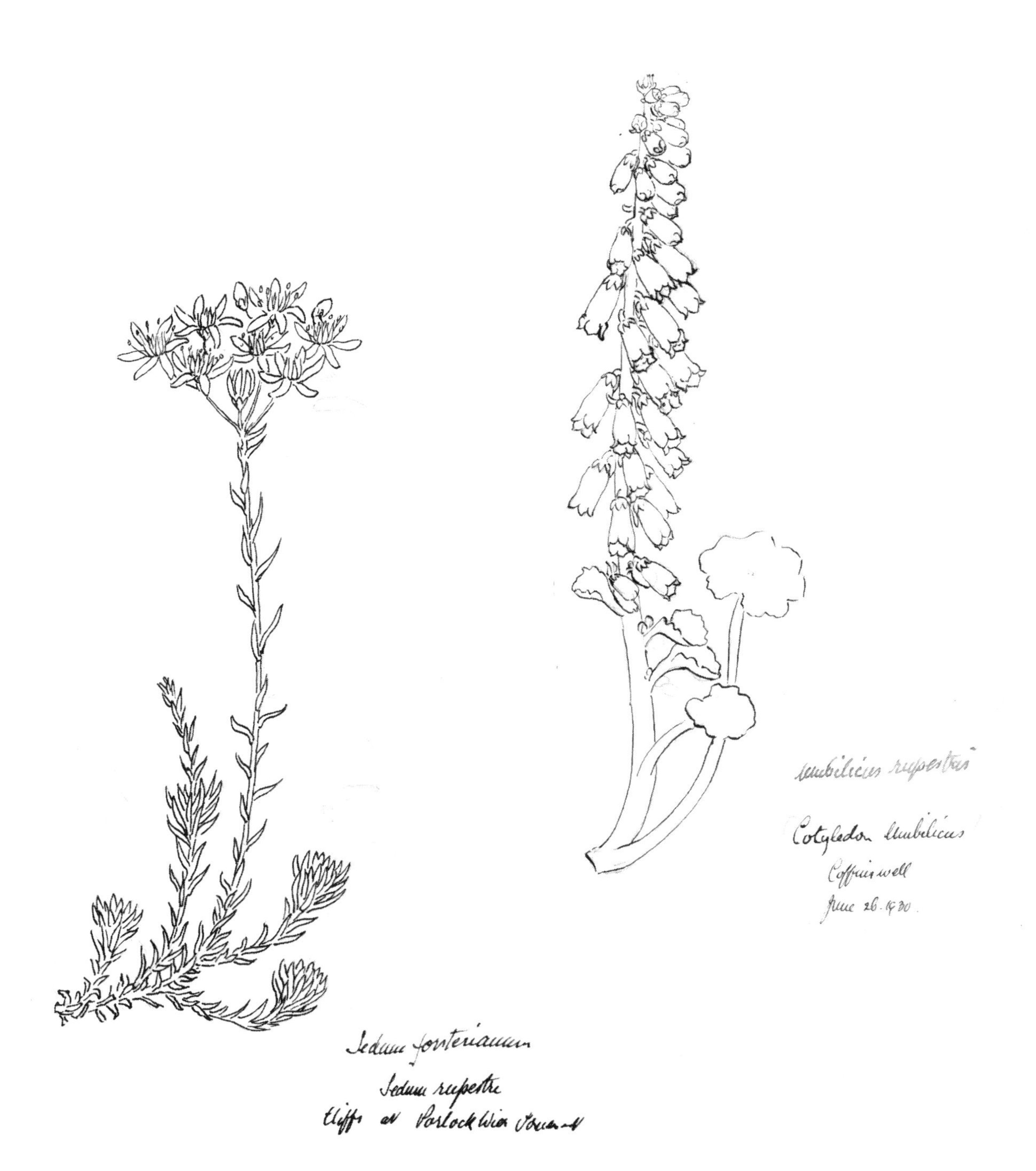

Sedum forsteranum Sm.
Rock Stonecrop

Umbilicus rupestris (Salisb.) Dandy.
Pennywort

Plate 34 LYTHRACEAE and HALORAGACEAE

Lythrum salicaria L.
Purple Loosestrife

Myriophyllum verticillatum L.
Whorled Water Milfoil

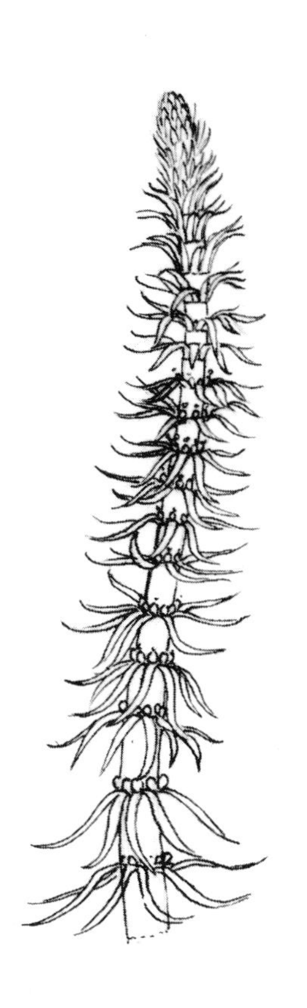

Drosera rotundifolia
open flowers in Walkham basin Nr Teignhead.
Aug. 20. 1926 11 a.m. in hot sunshine.

Hippuris vulgaris L
Godstow Lock, Oxfordshire
June 27th 1929
per Dr. G.C. Druce.

Hippuris vulgaris L.
Mare's Tail

Drosera rotundifolia L.
Sundew

Drosera intermedia Hayne.
Long-leaved Sundew

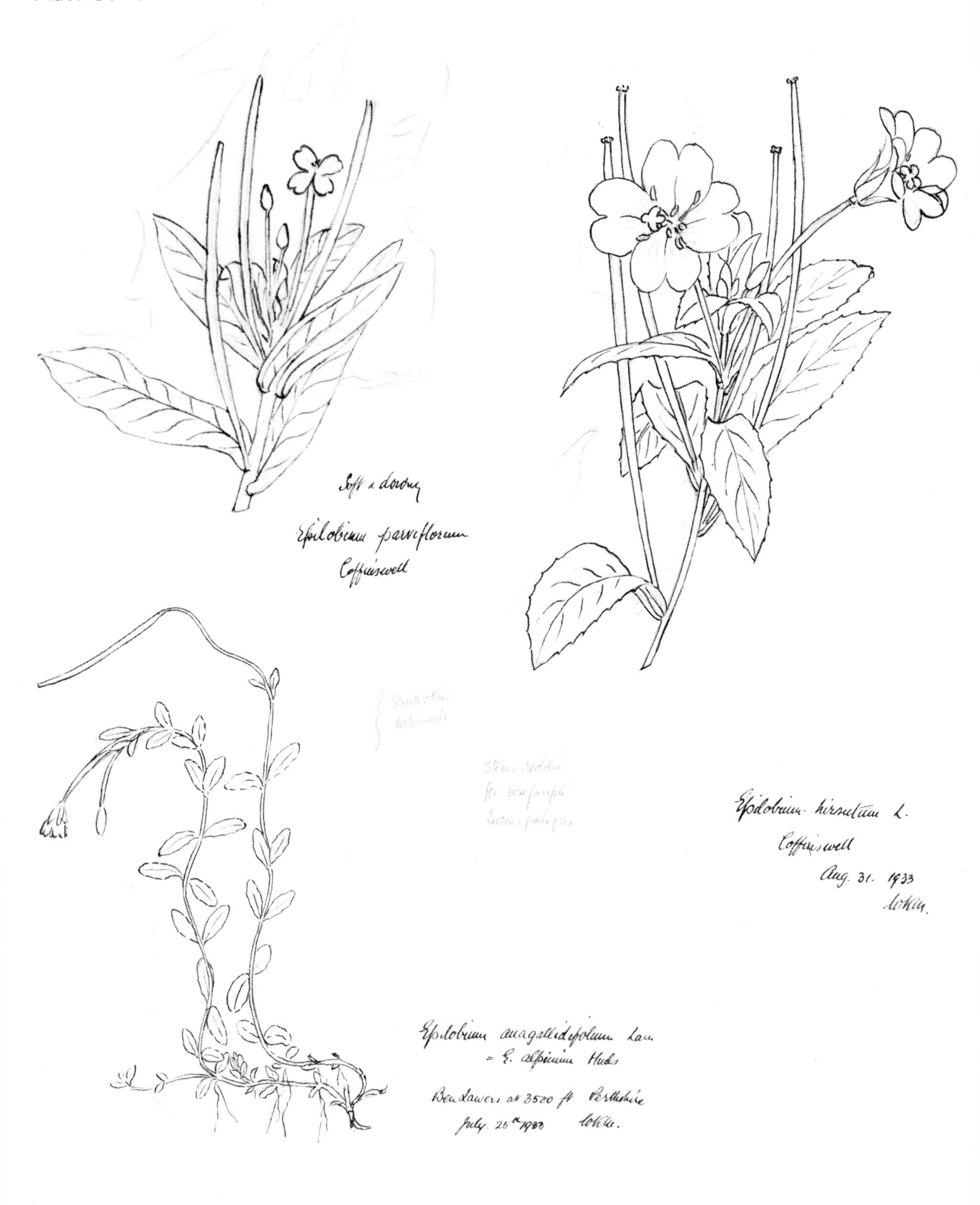

Epilobium parviflorum Schreb.
Hairy Willow Herb.

Epilobium anagallidifolium Lam.
Alpine Willow Herb

Epilobium hirsutum L.
Great Willow Herb

Epilobium angustifolium L.
Rose Bay

Epilobium roseum Schreb.
Pedicelled Willow Herb

Circaea alpina L.
Alpine Enchanter's Nightshade

Bryonia dioica
Nr Woodham Lodge
Addlestone, Surrey

Eryngium maritimum
Torcross Beach S. Devon
painted Aug 1st 1927 July 28 1927

Bryonia dioica Jacq.
White Bryony

Eryngium maritimum L.
Sea Holly

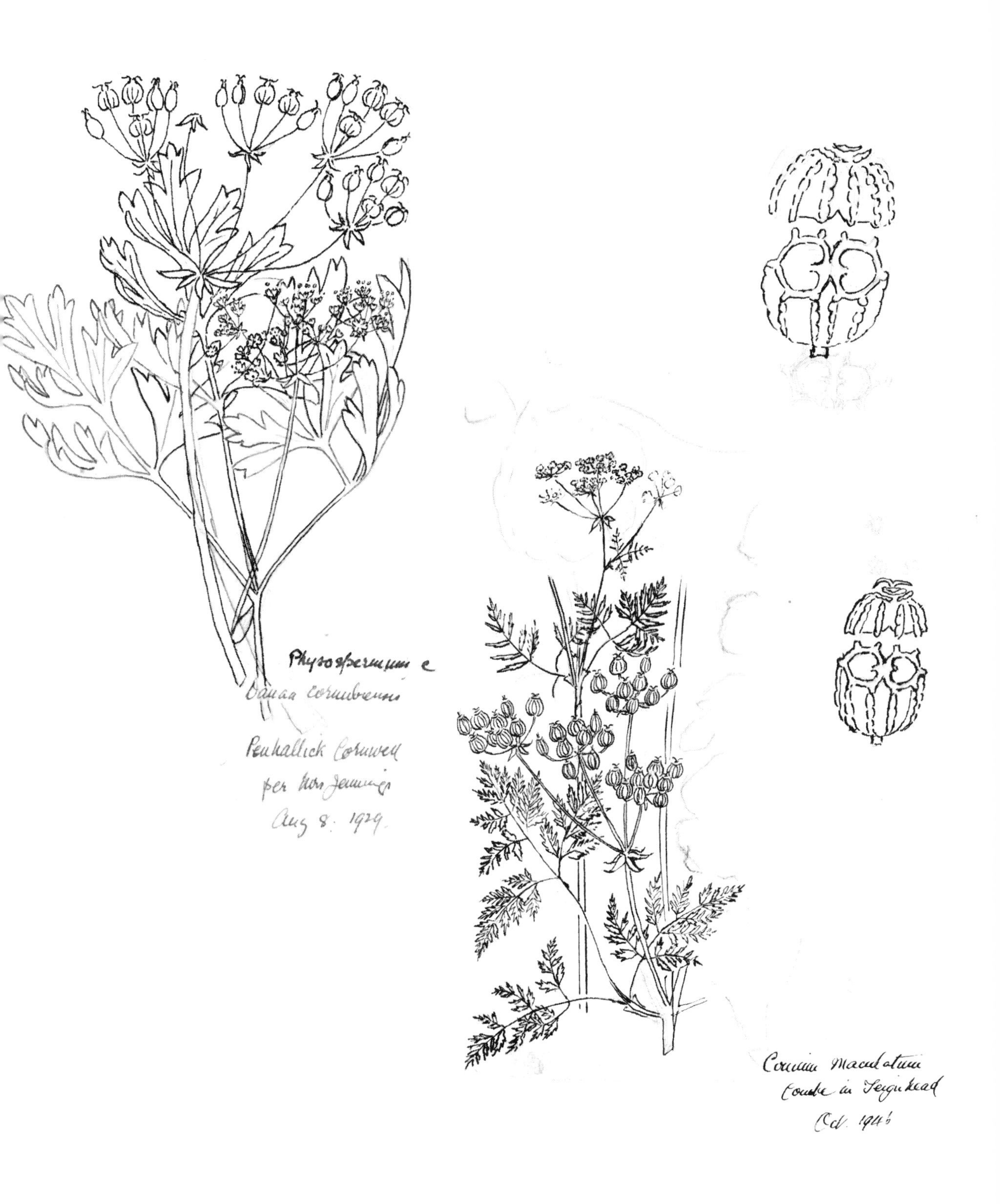

Physopermum cornubiense (L.) DC.
Bladderseed

Conium maculatum L.
Hemlock

Sium latifolium L.
Water Parsnip

Apium nodiflorum (L.) Lag.
Procumbent Marsh-wort

Bunium bulbocastanum L.
Tuberous Caraway

Cicuta virosa L.
Cowbane

Myrrhis odorata
riverside upper Teesdale
June 26. 1928. & July 1945
redrawn Nov. 1946

Myrrhis odorata (L.) Scop.
Sweet Cicely

Aegopodium podagraria L.
Goutweed, Ground Elder

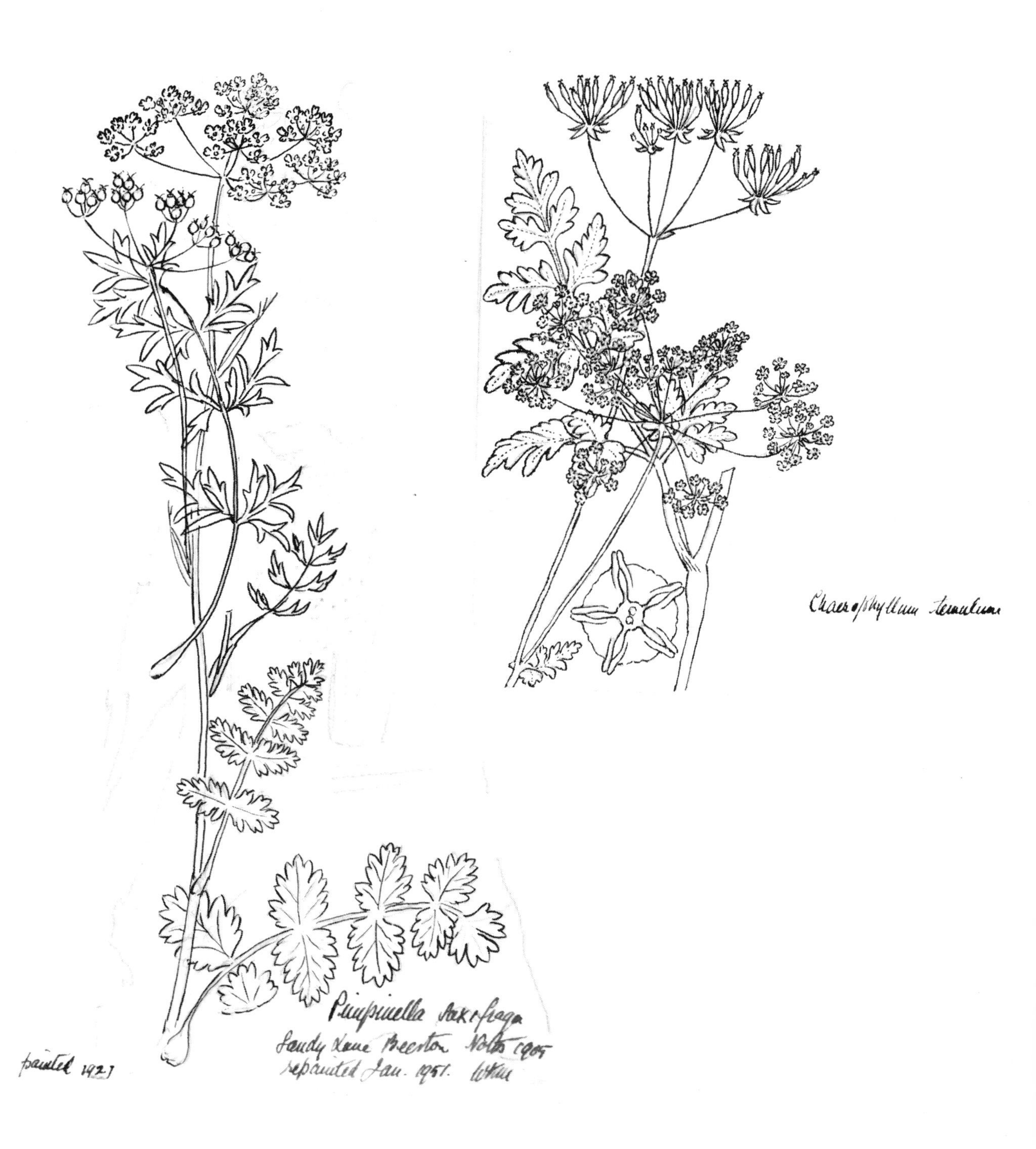

Pimpinella saxifraga L.
Burnet Saxifrage

Chaerophyllum temulentum L.
Chervil

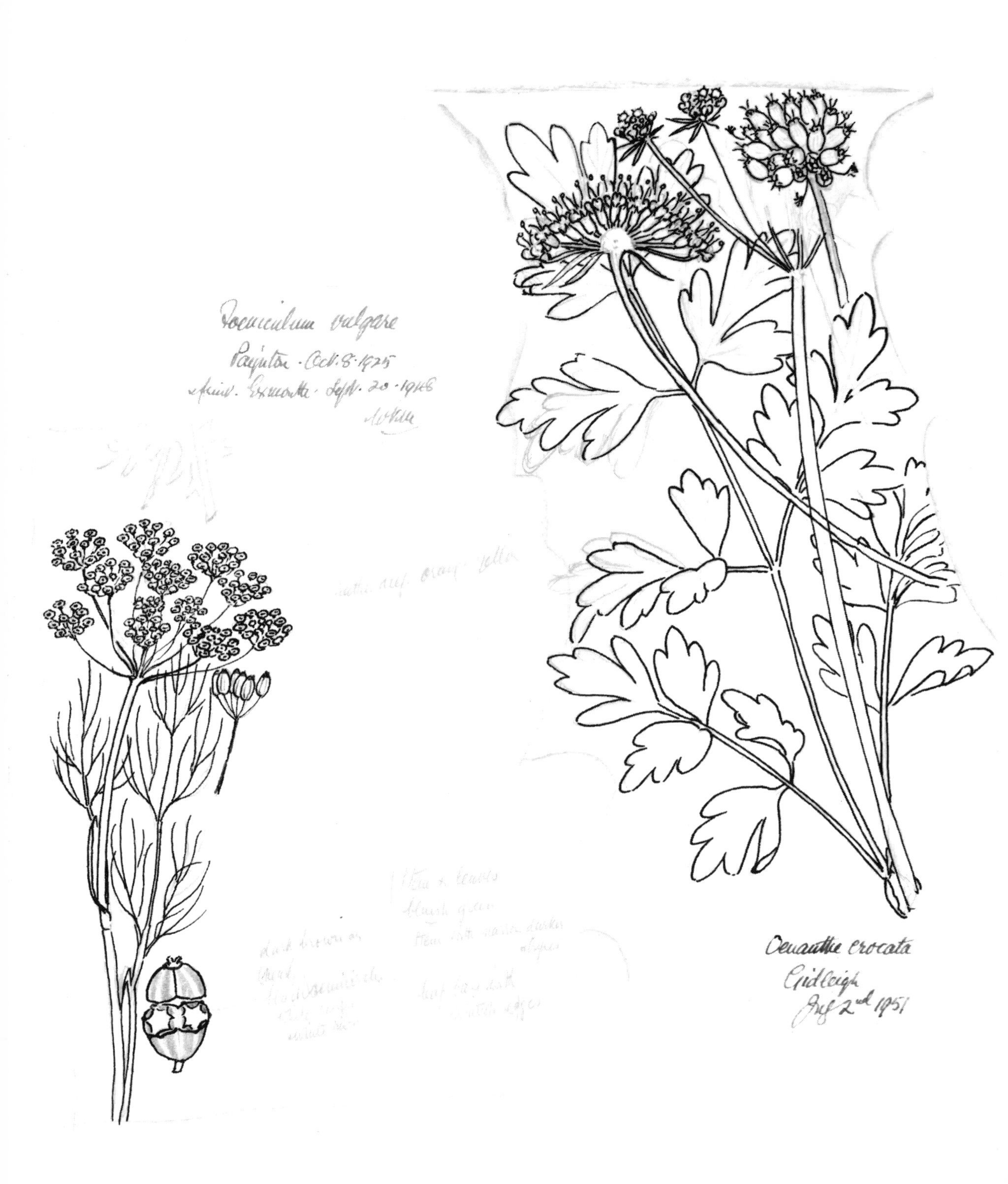

Foeniculum vulgare Mill.
Fennel

Oenanthe crocata L.
Hemlock Water Dropwort

Oenanthe pimpinelloides L.
Corky-fruited Water Dropwort

Redrawn from original sketches for the *Flora*

Ligusticum scoticum L.
Lovage

Torilis arvensis (Huds.) Link.
Spreading Hedge Parsley

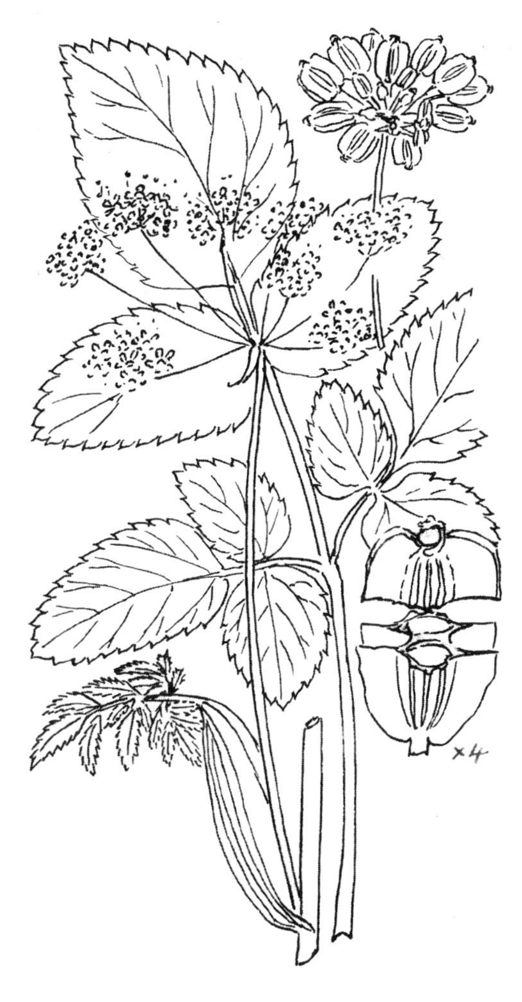

Angelica sylvestris
Gidleigh.
Sept. 25th 1950.

Heracleum sphondylium L.
Hogweed, Cow Parsnip

Angelica sylvestris L.
Wild Angelica

Swida sanguinea (L.) Opiz.
Dogwood, Cornel

Viburnum opulus L.
Guelder Rose

Linnaea borealis L.

Sambucus ebulus L.
Danewort

Galium saxatile = hercynicum
Gidleigh Dartmoor 1200 ft
July 12–26 1951

Galium odoratum
Asperula odorata

Sherardia arvensis
Combe in Teignhead
Oct. 8. 1945

Galium debile.
Chudleigh Knighton Heath, Devon

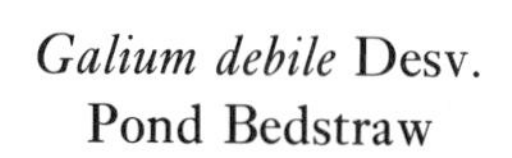

Galium odoratum (L.) Scop.
Sweet Woodruff

Galium saxatile L.
Heath Bedstraw

Galium debile Desv.
Pond Bedstraw

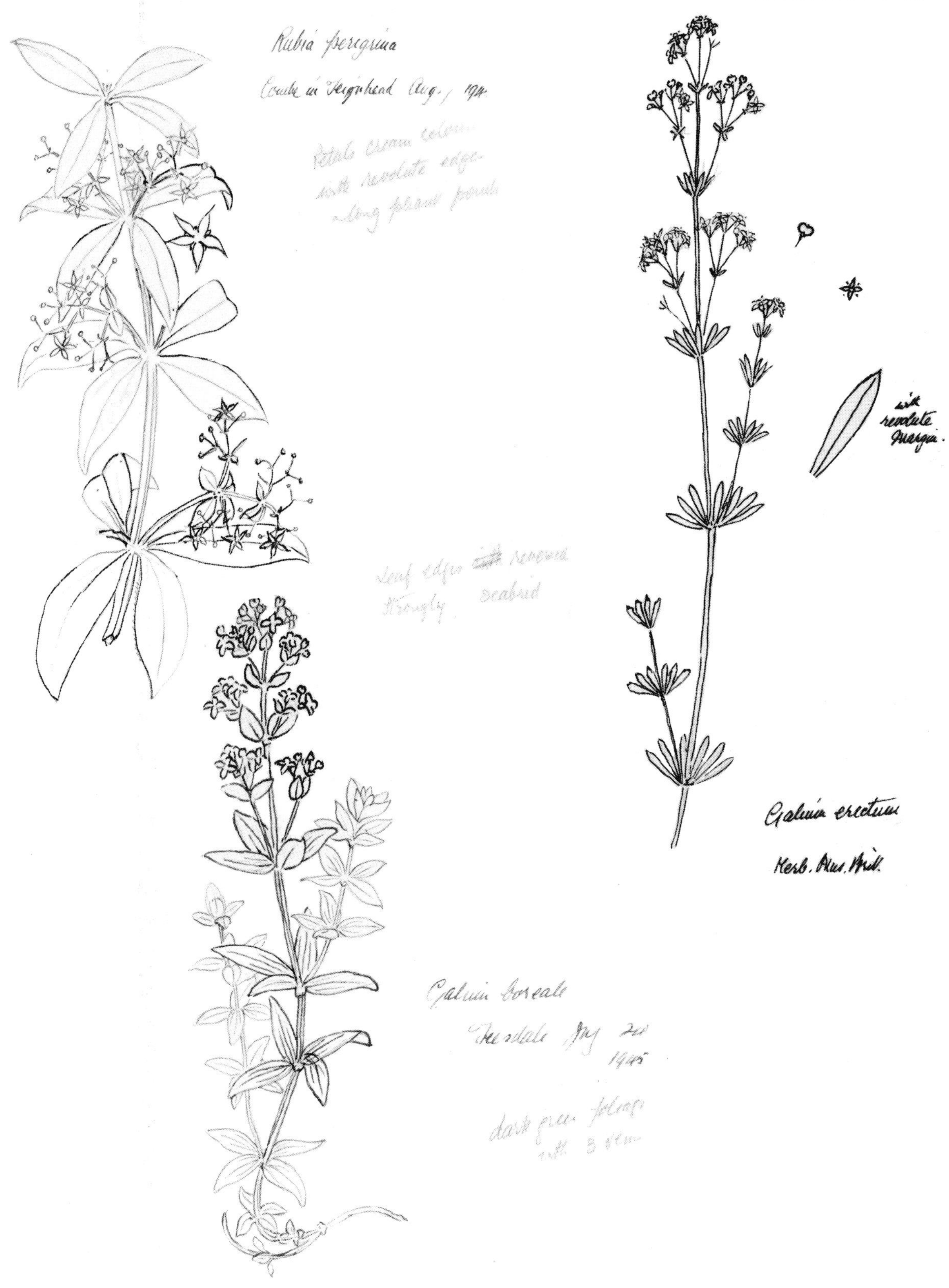

Rubia peregrina L.
Madder

Galium boreale L.
Northern Bedstraw

Galium album Mill.
Hedge Bedstraw

Valeriana dioica L.
Lesser Valerian

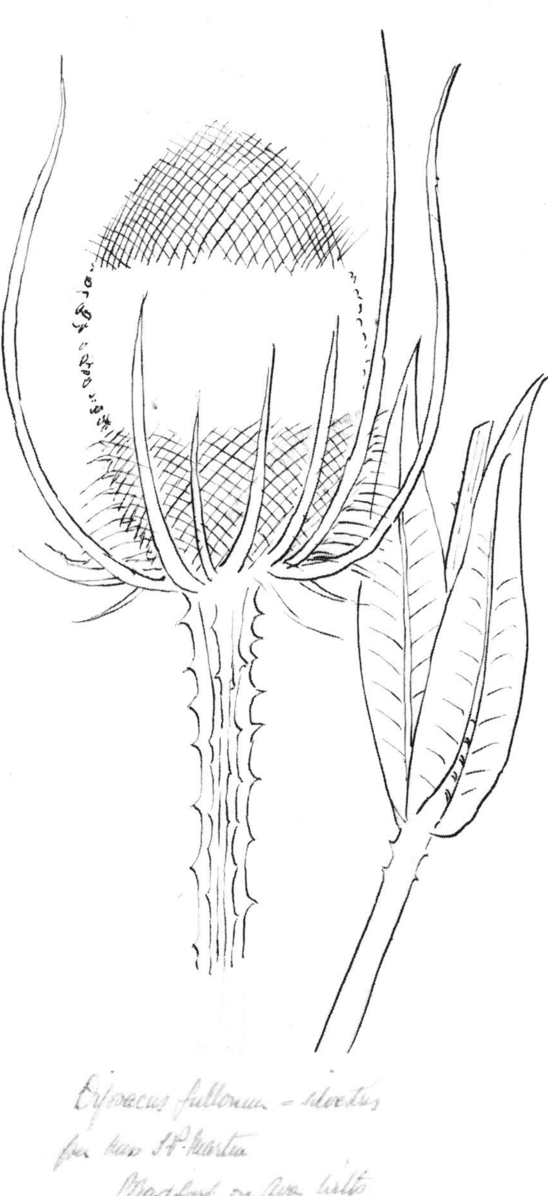

Dipsacus fullonum L.
Common Teasel

Valeriana officinalis L.
Valerian

Dipsacus pilosus L.
Small Teasel

Plate 44 COMPOSITAE

Crinitaria linosyris (L.) Less.
Goldilocks

Eupatorium cannabinum L.
Hemp Agrimony

Aster tripolium

Solidago Virgaurea

Aster tripolium L.
Sea Aster

Solidago virgaurea L.
Golden Rod

Antennaria dioica (L.) Gaertn.
Cat's foot

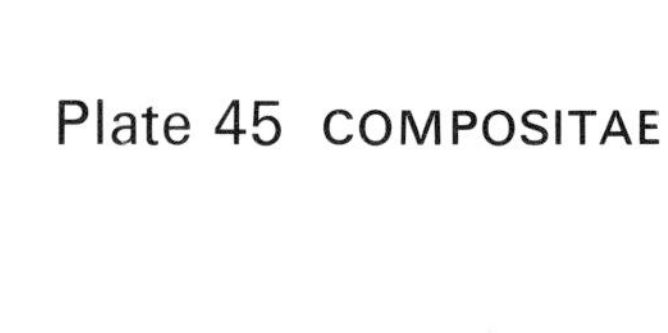

Pulicaria dysenterica [illegible]
Fleabane
Torrington Common
Oct. 1 1942

Pulicaria vulgaris Gaertn.
Lesser Fleabane

Pulicaria dysenterica (L.) Bernh.
Fleabane

Bidens tripartita L.
Bur-Marigold

Otanthus maritimus (L.) Hoffmanns. & Link.

Inula helenium L.
Elecampane.

Tripleurospermum maritimum (L.) Koch
Scentless Chamomile

Matricaria matricarioides (Less.) Porter.
Rayless Mayweed

Artemisia maritima L.
Sea Wormwood

Artemisia vulgaris L.
Mugwort

Artemisia campestris L.
Breckland Wormwood

Artemisia absinthium L.
Wormwood

Plate 47 COMPOSITAE

Senecio sylvaticus L.
Wood Groundsel

Arctium pubens Bab.
Common Burdock

Senecio erucifolius L.
Hoary Ragwort

Senecio spathulifolius Turcz.
Spathulate Fleawort

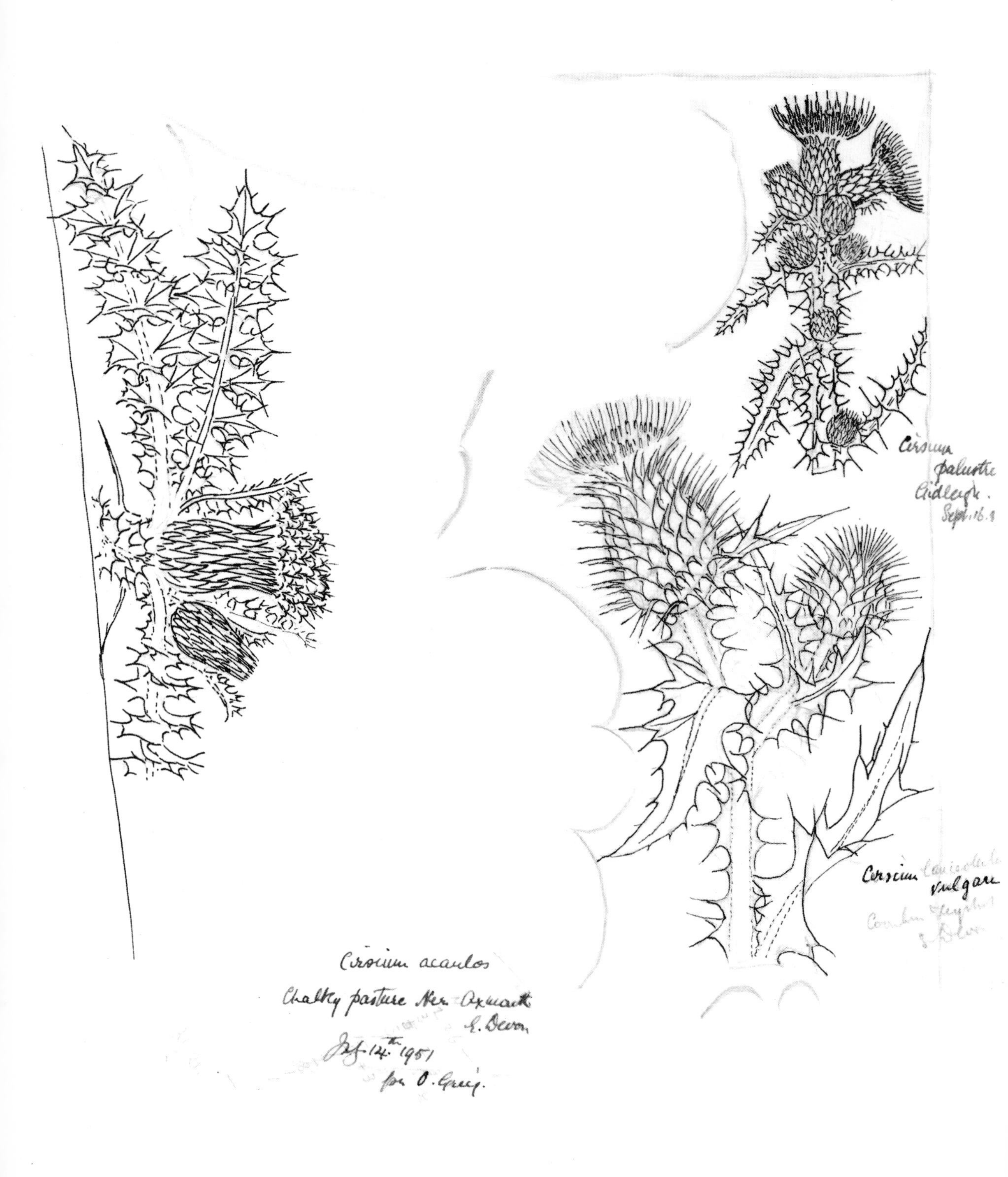

Cirsium acaule Scop.
Ground Thistle

Cirsium palustre (L.) Scop.
Marsh Thistle

Cirsium vulgare (Savi.) Ten.
Spear Thistle

Carduus nutans L.
Nodding Thistle

Carduus tenuiflorus Curt.
Slender-flowered Thistle

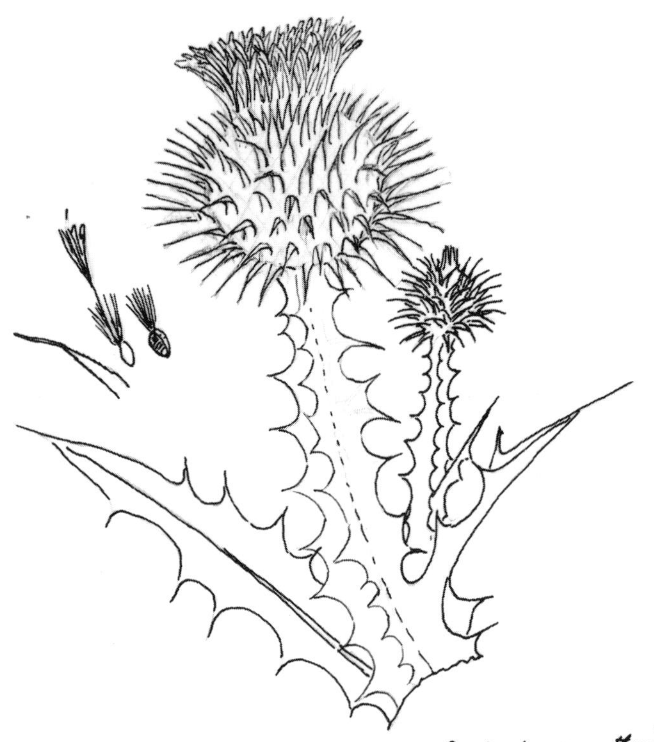

Onopordon acanthium

Nr Brandon, Suffolk. Aug 20. 1932
and Woodham Lodge Addlestone, Surrey

Serratula tinctoria
Winch Br. Teesdale
Jy 25. paint Aug. 2. 1945

Onopordon acanthium L.
Scottish Thistle

Serratula tinctoria L.
Saw-wort

Centaurea aspera L.

Centaurea scabiosa L.
Great Knapweed

Centaurea calcitrapa L.
Star Thistle

Plate 50 COMPOSITAE

Crepis foetida L. Stinking Hawk's-beard	*Crepis mollis* (Jacq.) Aschers. Soft Hawk's-beard	*Picris hieracioides* L. Hawkweed Ox-tongue

Lapsana communis L.
Nipple-wort

Crepis biennis L.
Greater or Rough Hawk's-beard

Plate 51 COMPOSITAE

Redrawn from original sketches for the *Flora*

Redrawn from original sketches for the *Flora*

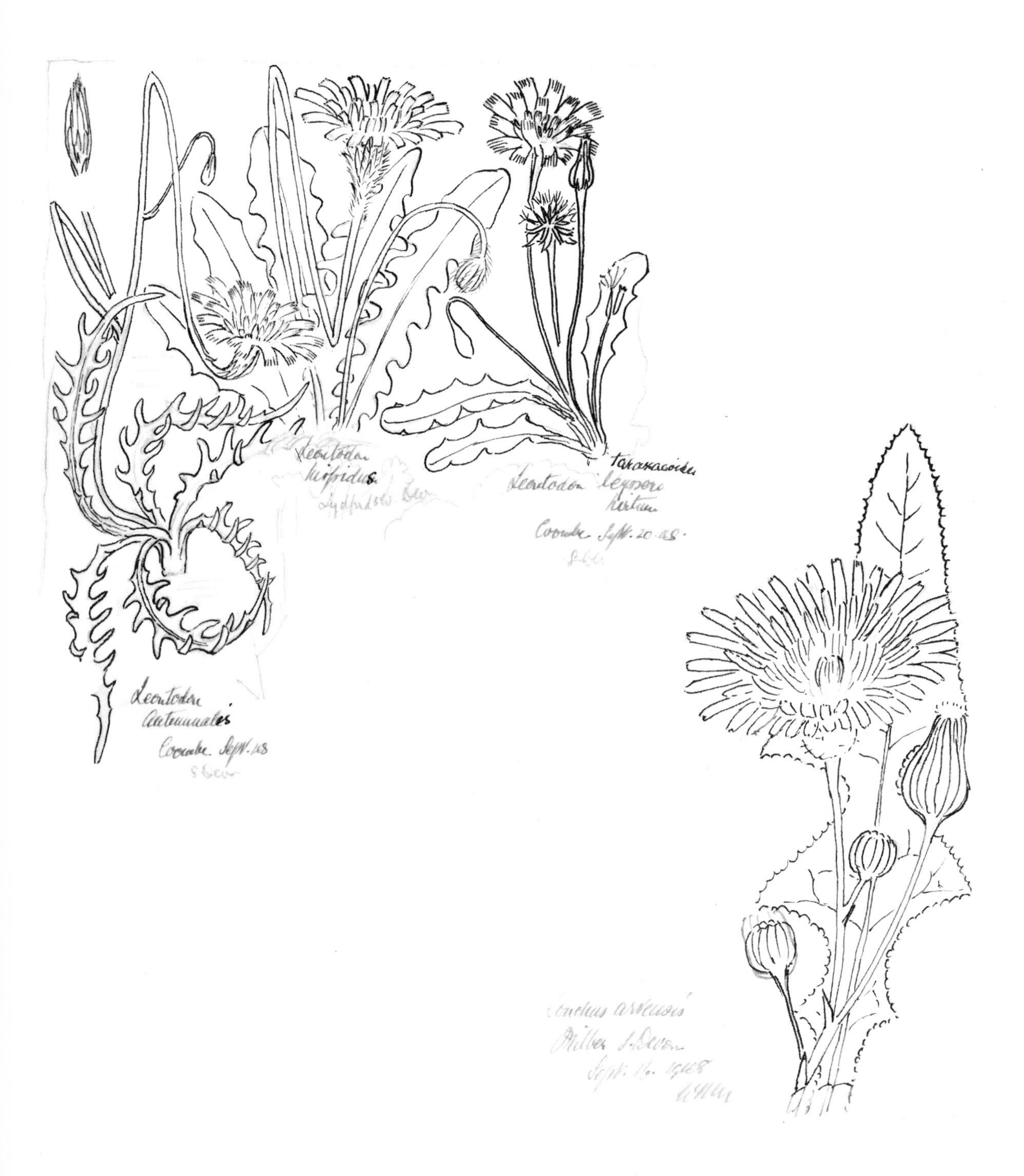

Leontodon autumnalis L.
Smooth Hawkbit

Leontodon hispidus L.
Rough Hawkbit

Leontodon taraxacoides (Vill.) Merat.
Hawkbit

Sonchus arvensis L.
Corn Sowthistle

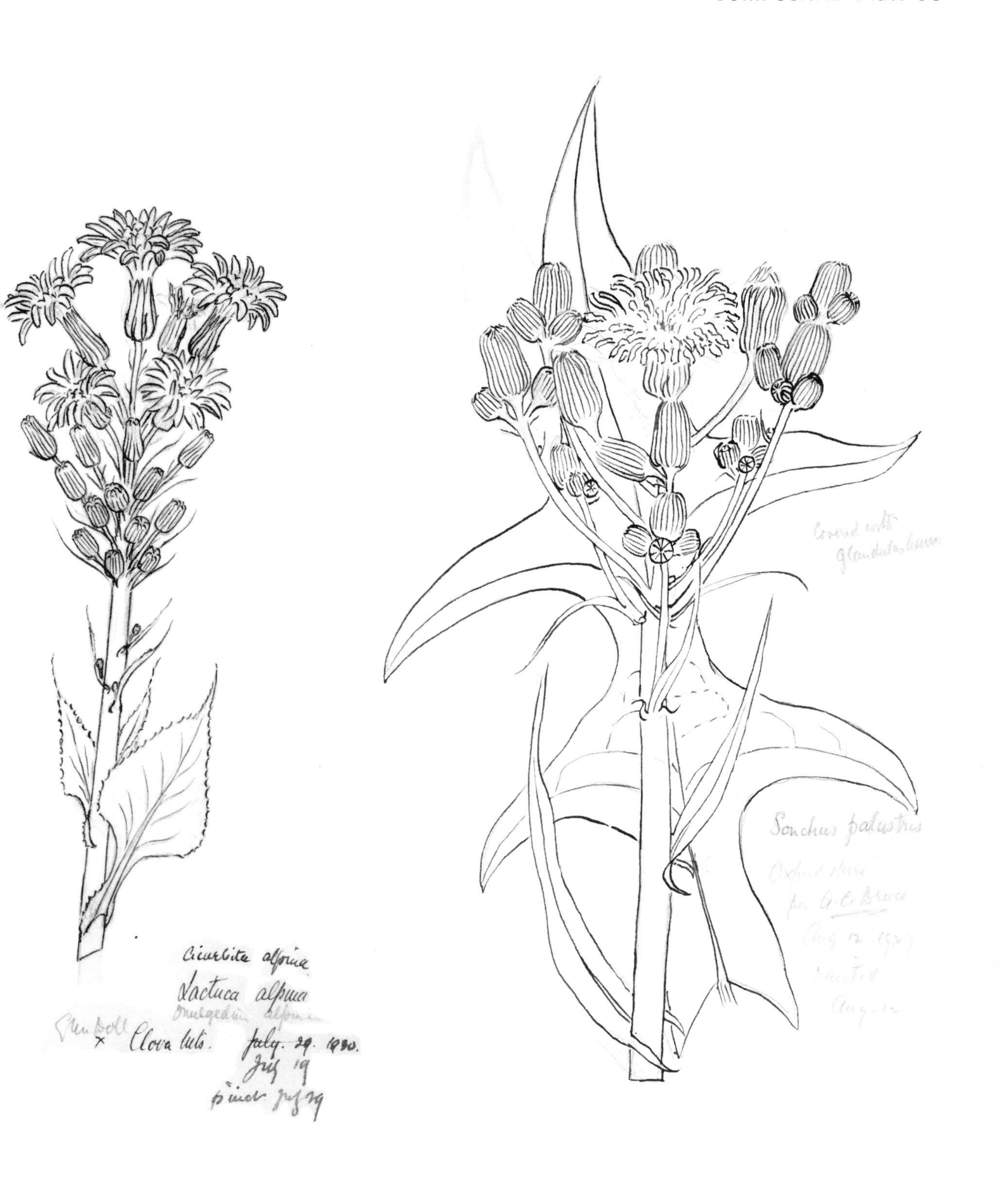

Cicerbita alpina (L.) Wallr.
Alpine Lettuce

Sonchus palustris L.
Fen Sowthistle

Lobelia urens L.
Blue Lobelia

Campanula patula L.
Spreading Bell-flower

Campanula glomerata
Stanway Cotswolds Glos.
Aug. 5 - 1912
W.K.H.

Redrawn Aug. 4. 1948

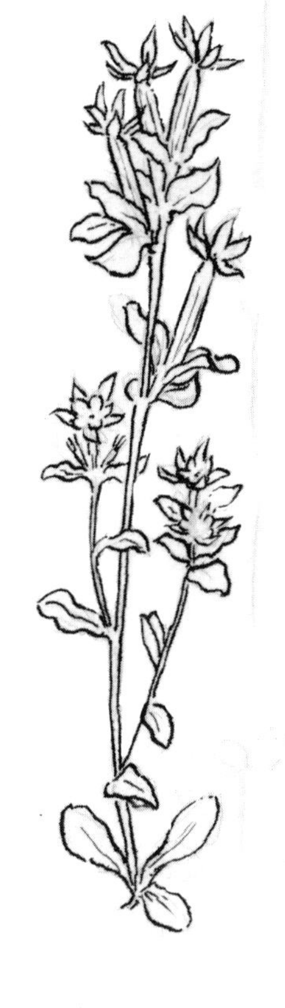

Specularia hybrida
per Dr R.W. [illegible], Bramcote, Notts
leg. June 30. drawn July 5. [illegible]

Campanula glomerata L.
Clustered Bell-flower

Legousia hybrida (L.) Delarb.
Venus' Looking-glass.

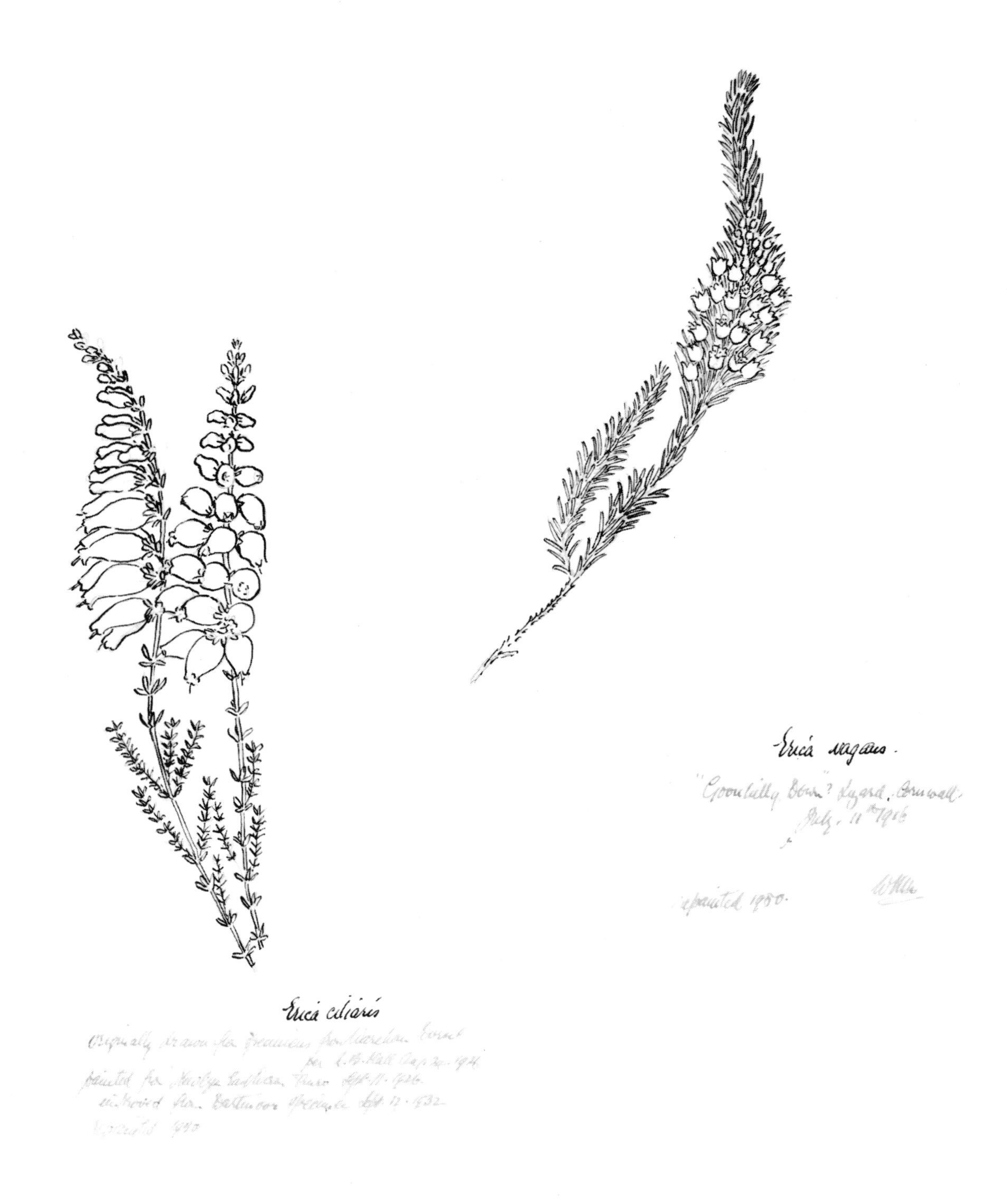

Erica ciliaris L.
Ciliate Heath

Erica vagans L.
Cornish Heath

Vaccinium myrtillus

Arctostaphylos uva-ursi

Loch Brandy, Clova hills,

2200 ft. June 15. 1931.

Vaccinium myrtillus L.
Bilberry, Whortleberry

Arctostaphylos uva-ursi (L.) Spreng.
Bear Berry

Arbutus unedo L.
Strawberry Tree

Pyrola rotundifolia L.
Large Wintergreen

Monenses uniflora (L.) A. Gray.
One-flowered Wintergreen

Limonium bellidifolium (Gouan) Dumort.

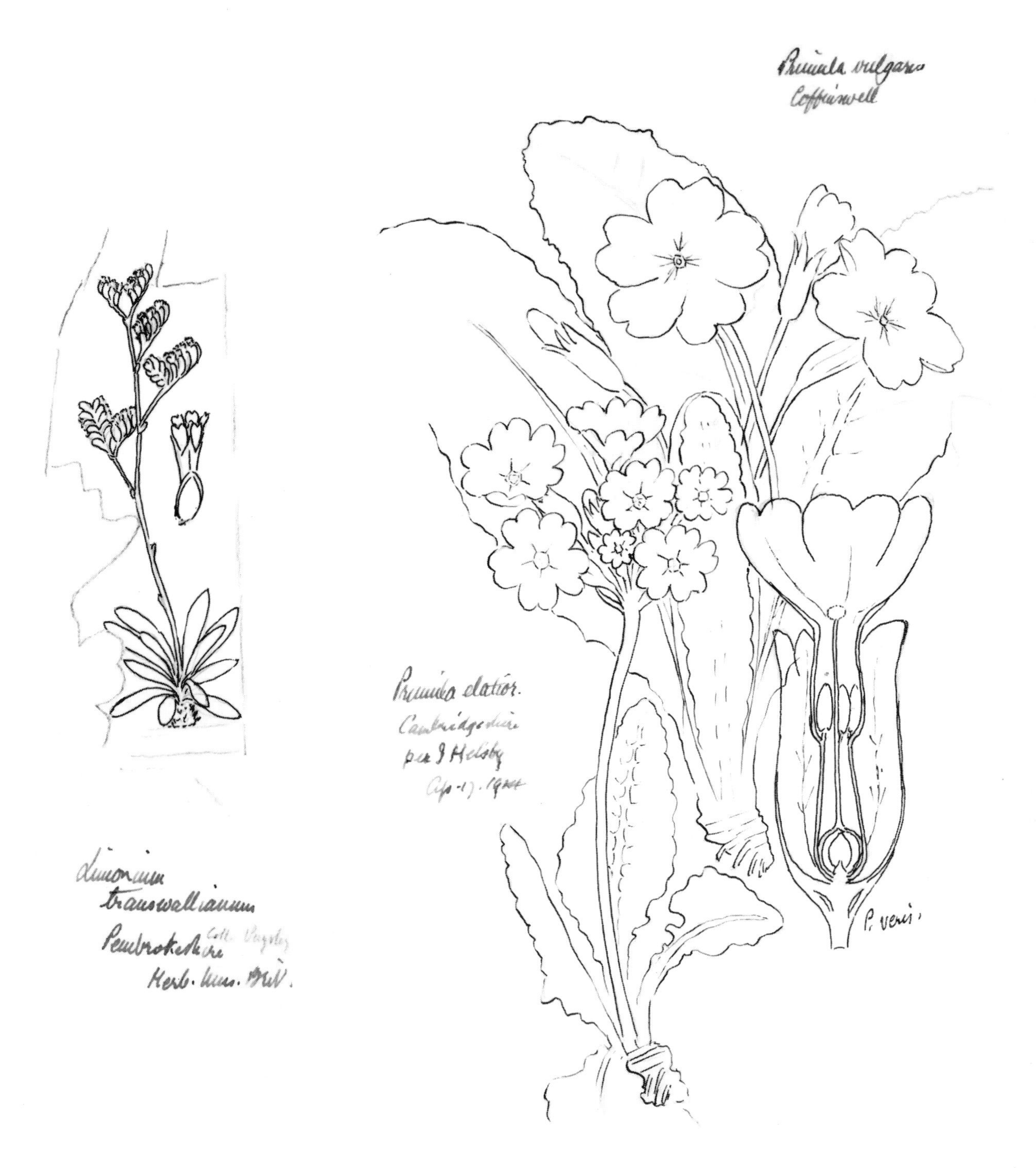

Limonium transwallianum (Pugsl.) Pugsl.

Primula elatior (L.) Hill.
Oxlip

Primula vulgaris Huds.
Primrose

Plate 57 PRIMULACEAE

Lysimachia vulgaris L.
Yellow Loosestrife

Lysimachia nemorum L.
Yellow Pimpernel

Lysimachia nummularia L.
Creeping Jenny

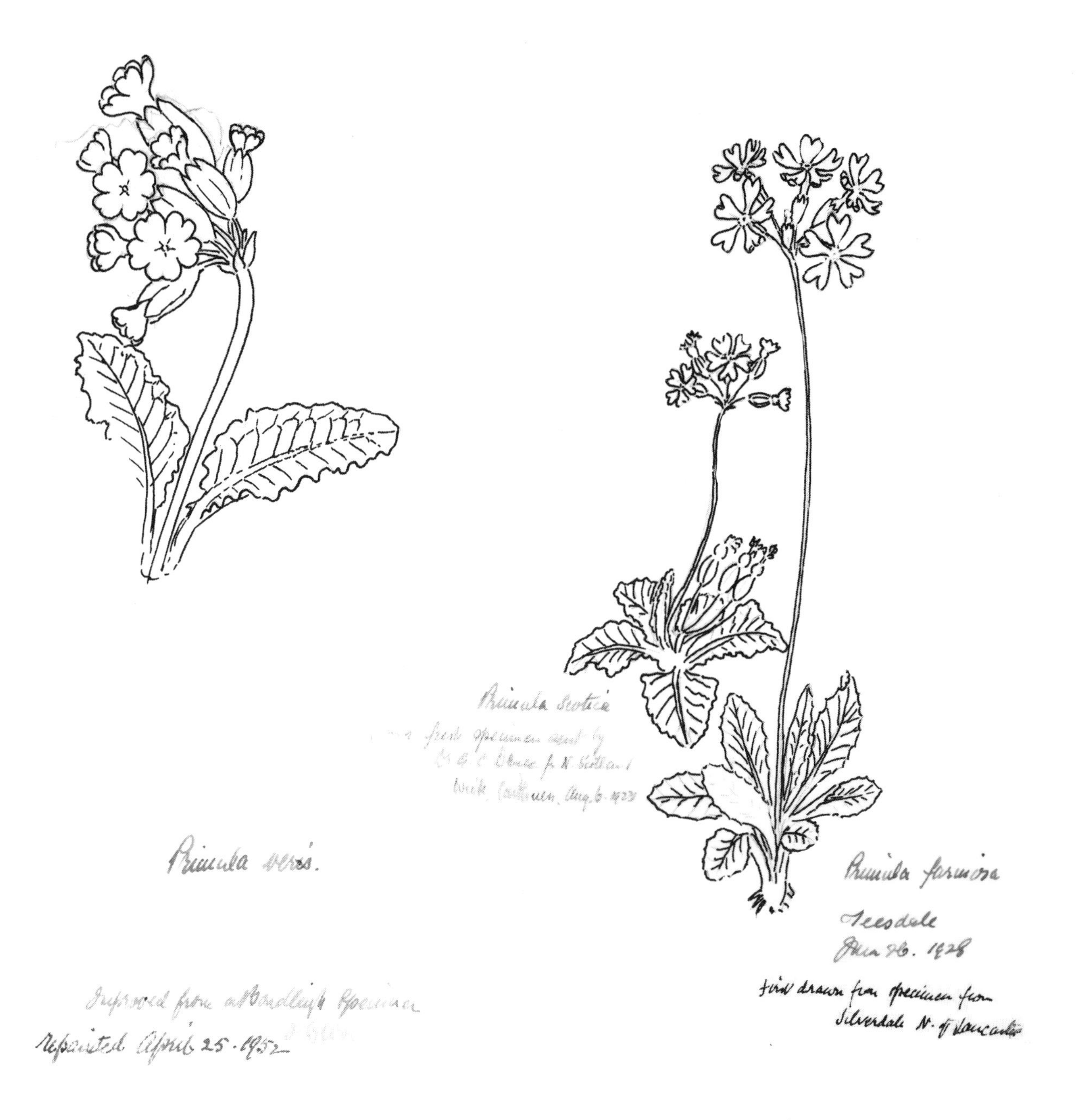

Primula veris L.
Cowslip

Primula scotica Hook.

Primula farinosa L.
Bird's-eye Primrose

Centaurium pulchellum (SW.) Druce.
Lesser Centaury

Centaurium erythraea Rafn.
Common Centaury

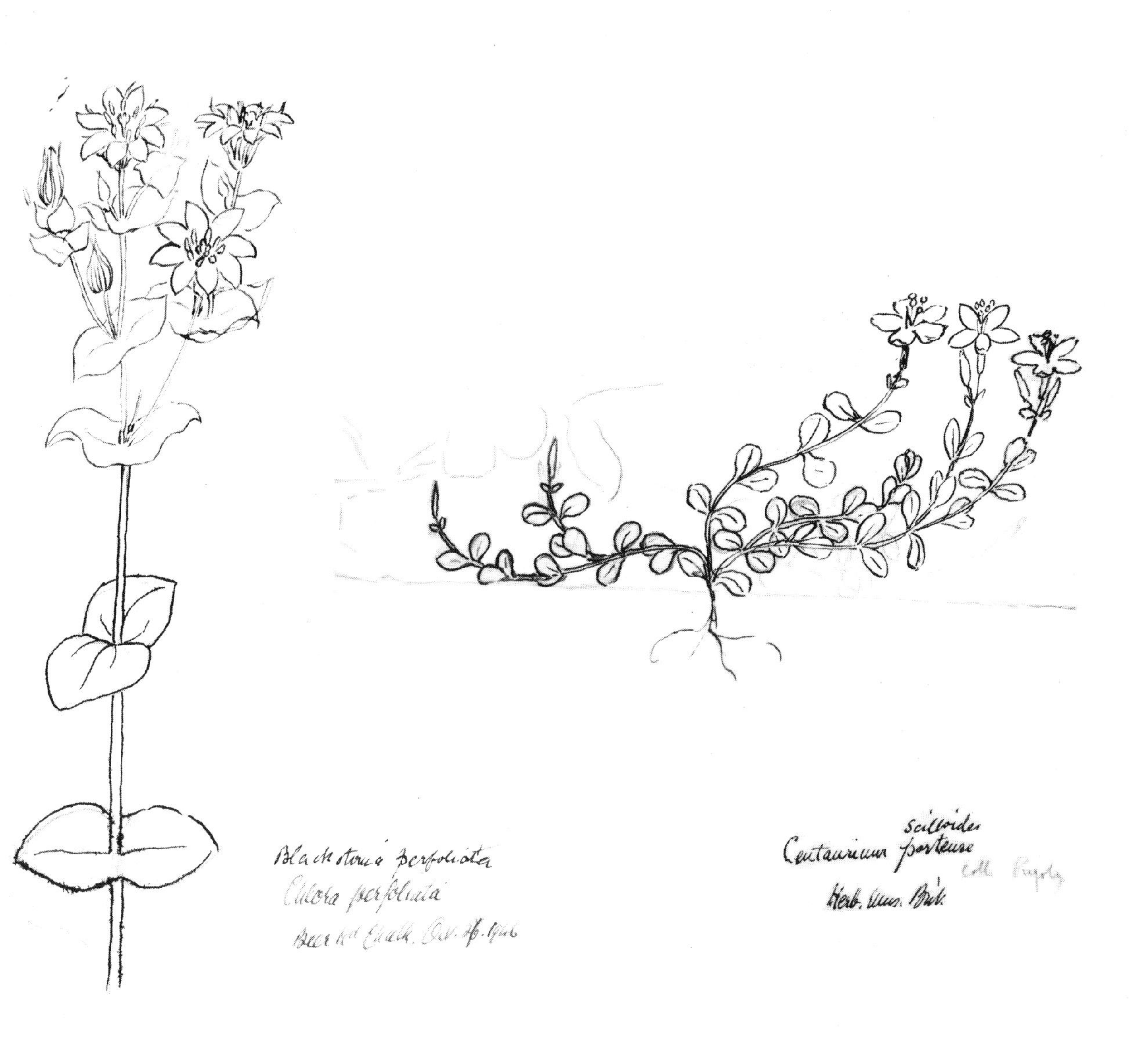

Blackstonia perfoliata (L.) Huds.
Yellow-wort

Centaurium scilloides (L.f.) Samp.
Perennial Centaury

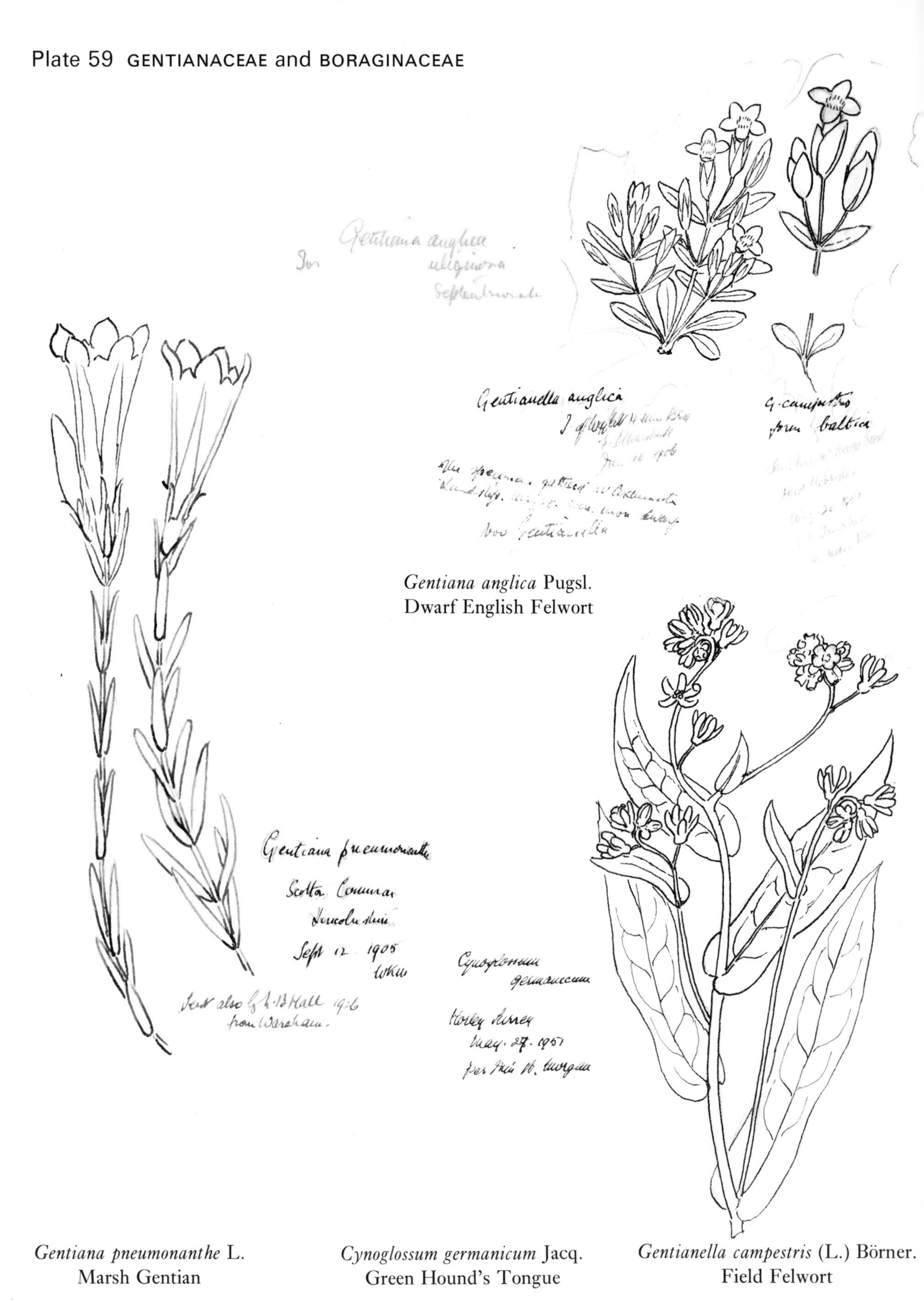

Gentiana anglica Pugsl.
Dwarf English Felwort

Gentiana pneumonanthe L.
Marsh Gentian

Cynoglossum germanicum Jacq.
Green Hound's Tongue

Gentianella campestris (L.) Börner.
Field Felwort

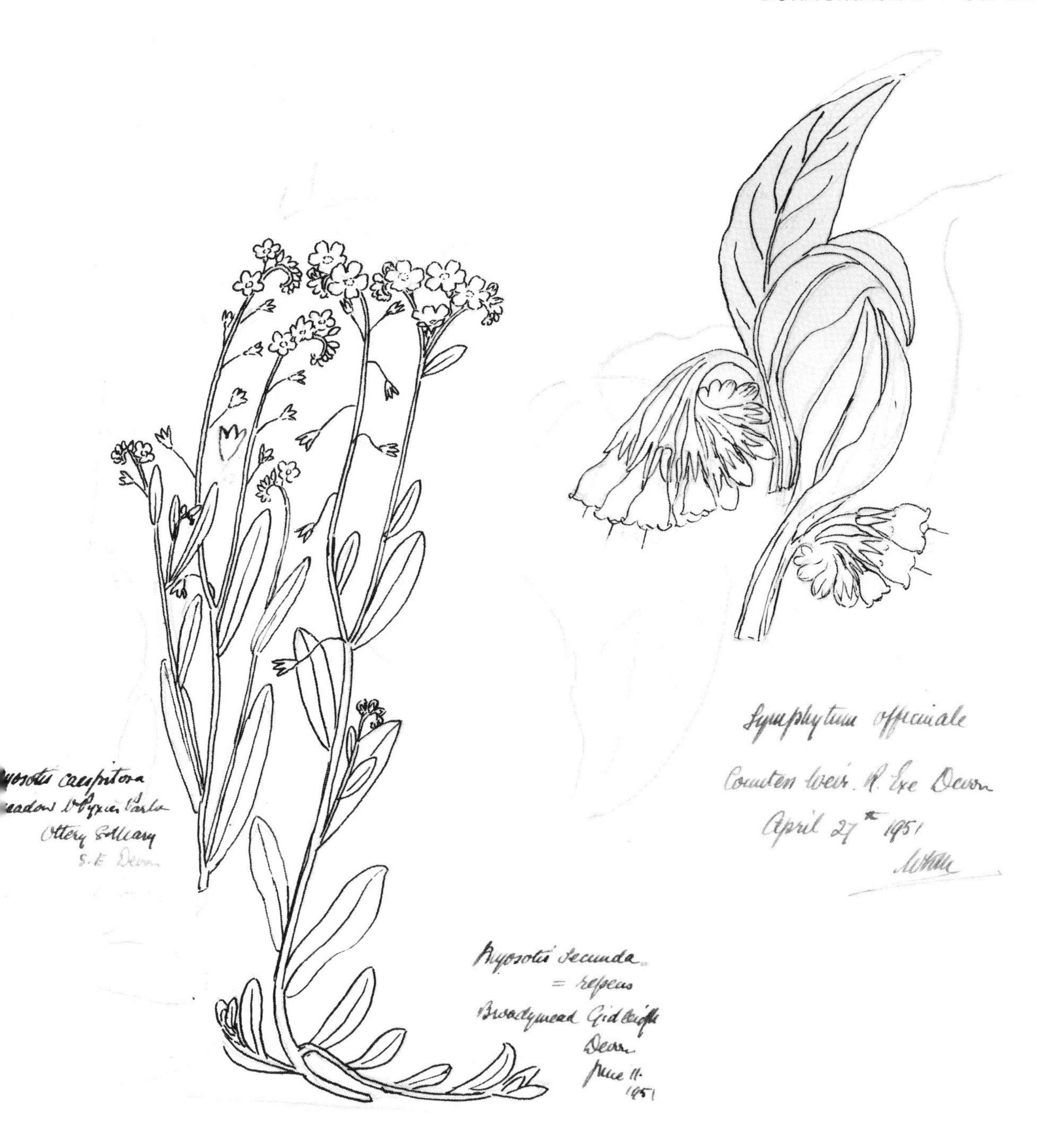

Myosotis caespitosa K. F. Schultz.	*Myosotis secunda* A. Murr.	*Symphytum officinale* L.
Lesser Water Forget-me-not	Marsh Forget-me-not	Comfrey

Symphytum tuberosum L.
Tuberous Comfrey

Pentaglottis sempervirens (L.) Tausch.
Alkanet

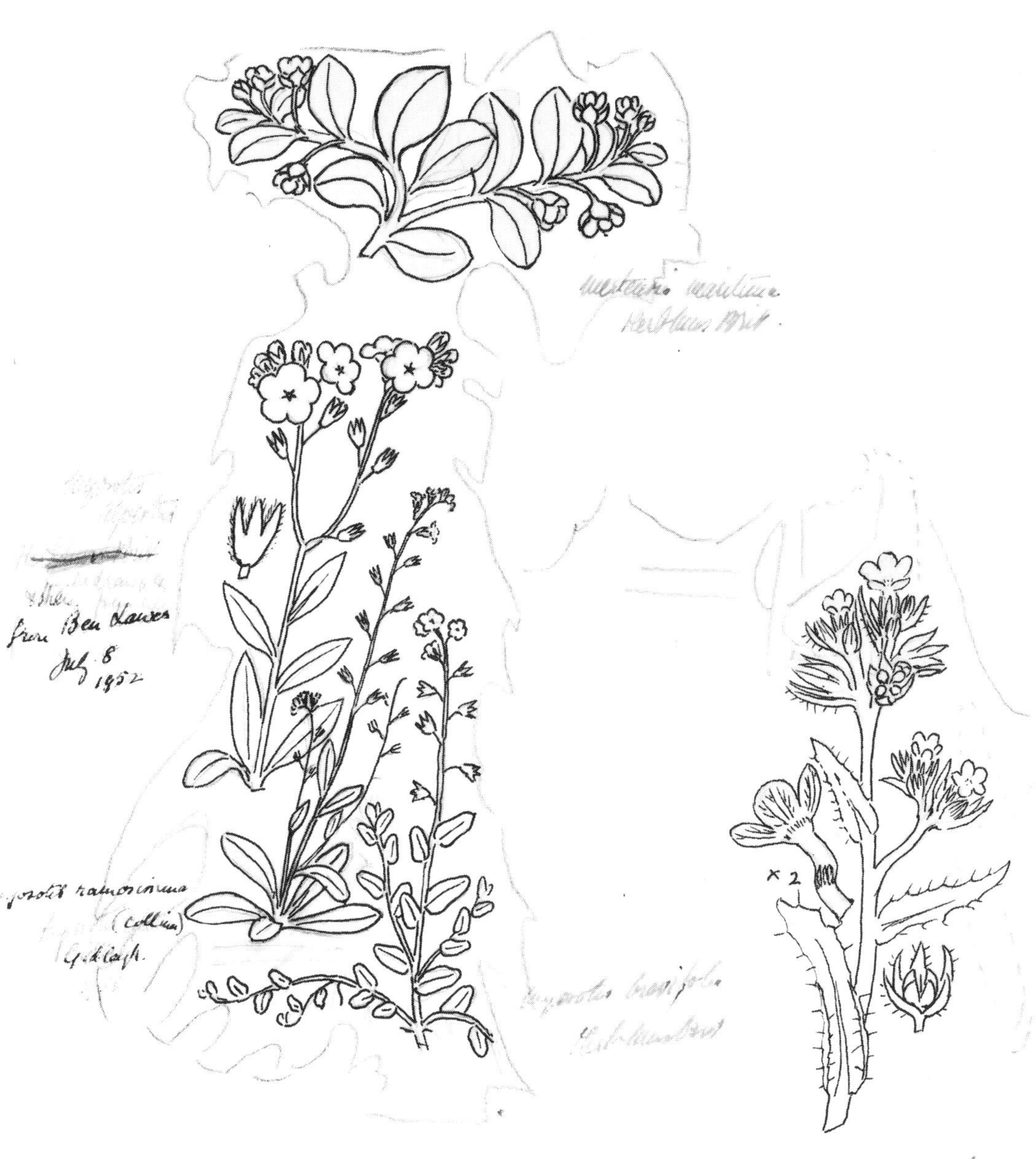

Myosotis alpestris Schmidt.
Alpine Forget-me-not

Myosotis ramosissima Rochel.
Early Forget-me-not

Mertensia maritima (L.) Gray.
Northern Shore-wort

Myosotis stolonifera Gay.
Short-leaved Forget-me-not

Lycopsis arvensis L.
Lesser Bugloss

Atropa bella-donna L.
Deadly Nightshade

Echium vulgare L.
Viper's Bugloss

Solanum nigrum L.
Black Nightshade

Solanum dulcamara L.
Woody Nightshade

Cuscuta europaea L.
Greater Dodder

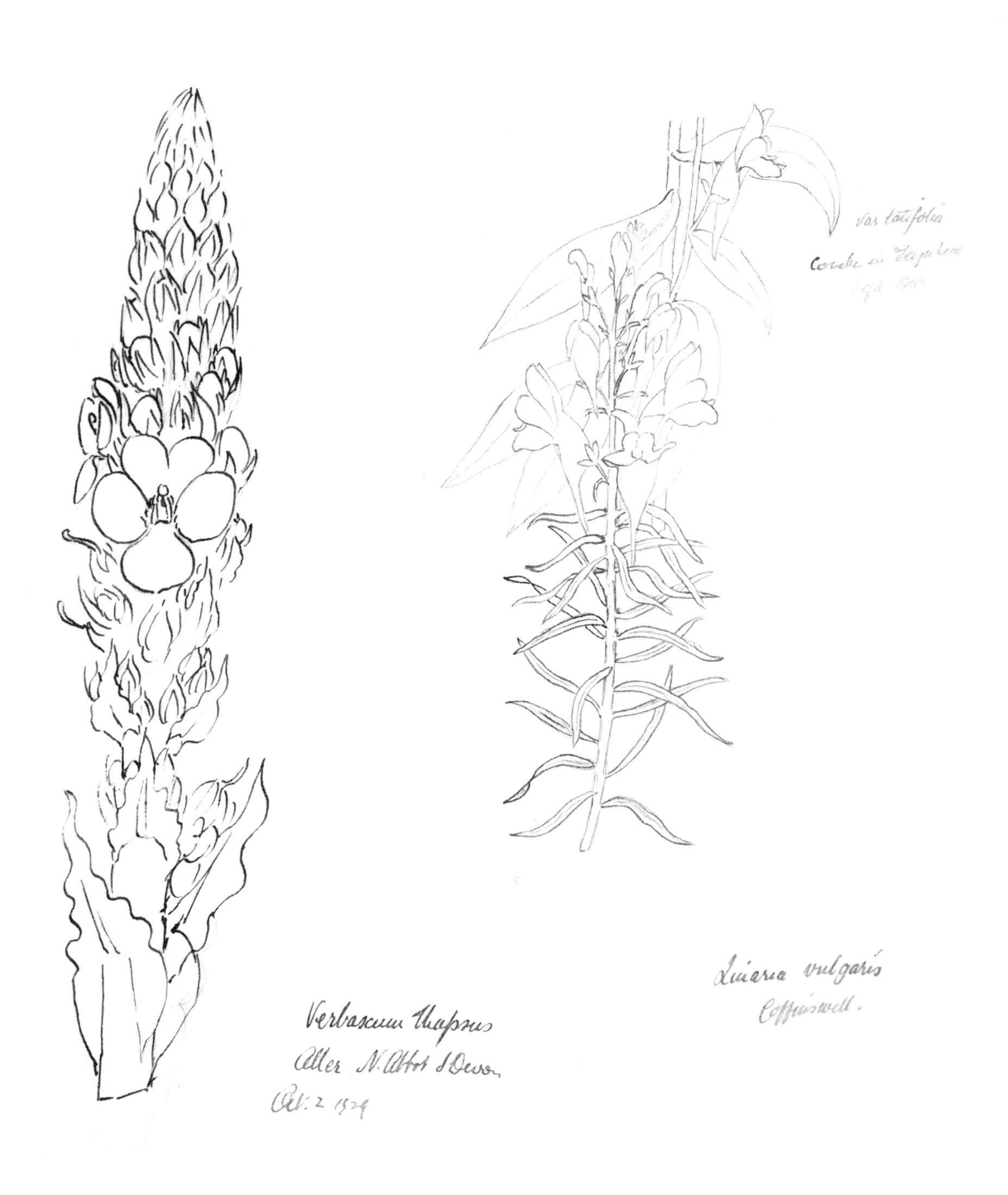

Verbascum thapsus L.
Common Mullein

Linaria vulgaris Mill.
Common Toadflax

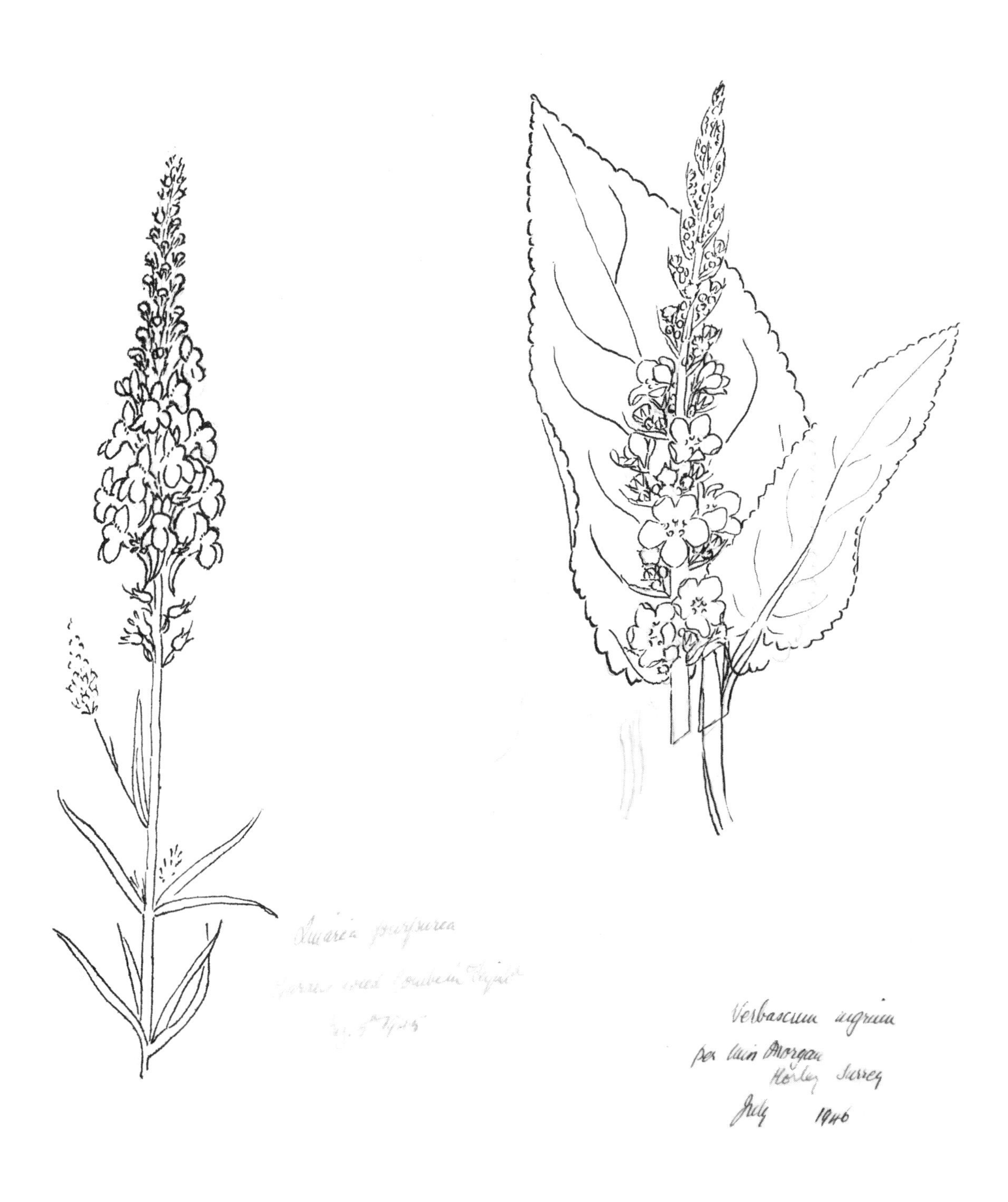

Linaria purpurea (L.) Mill
Purple Toadflax

Verbascum nigrum L.
Dark Mullein

Scrophularia Scorodonia

St Ives - June 6. 1929

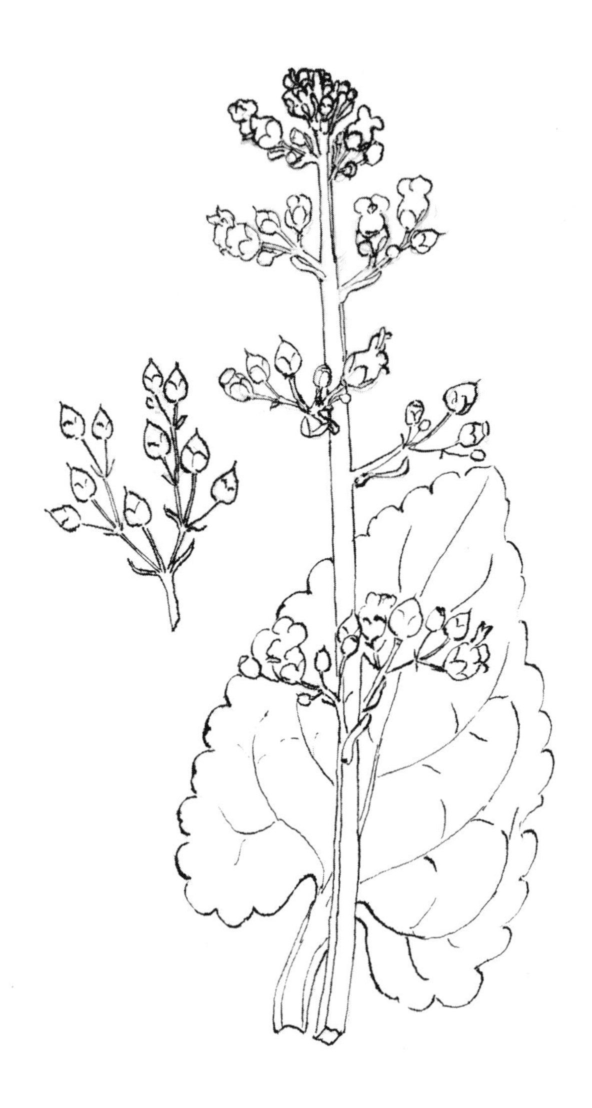

Scrophularia aquatica

Torrington Nov. 20 1942.

Scrophularia scorodonia L.
Balm-leaved Figwort

Scrophularia aquatica auct.
Water Figwort

Veronica polita Fr.
Grey Speedwell

Veronica persica Poir.
Persian Speedwell

Veronica verna L.
Spring Speedwell

Veronica spicata L.
Spiked Speedwell

Veronica fruticans Jacq.
Shrubby Speedwell

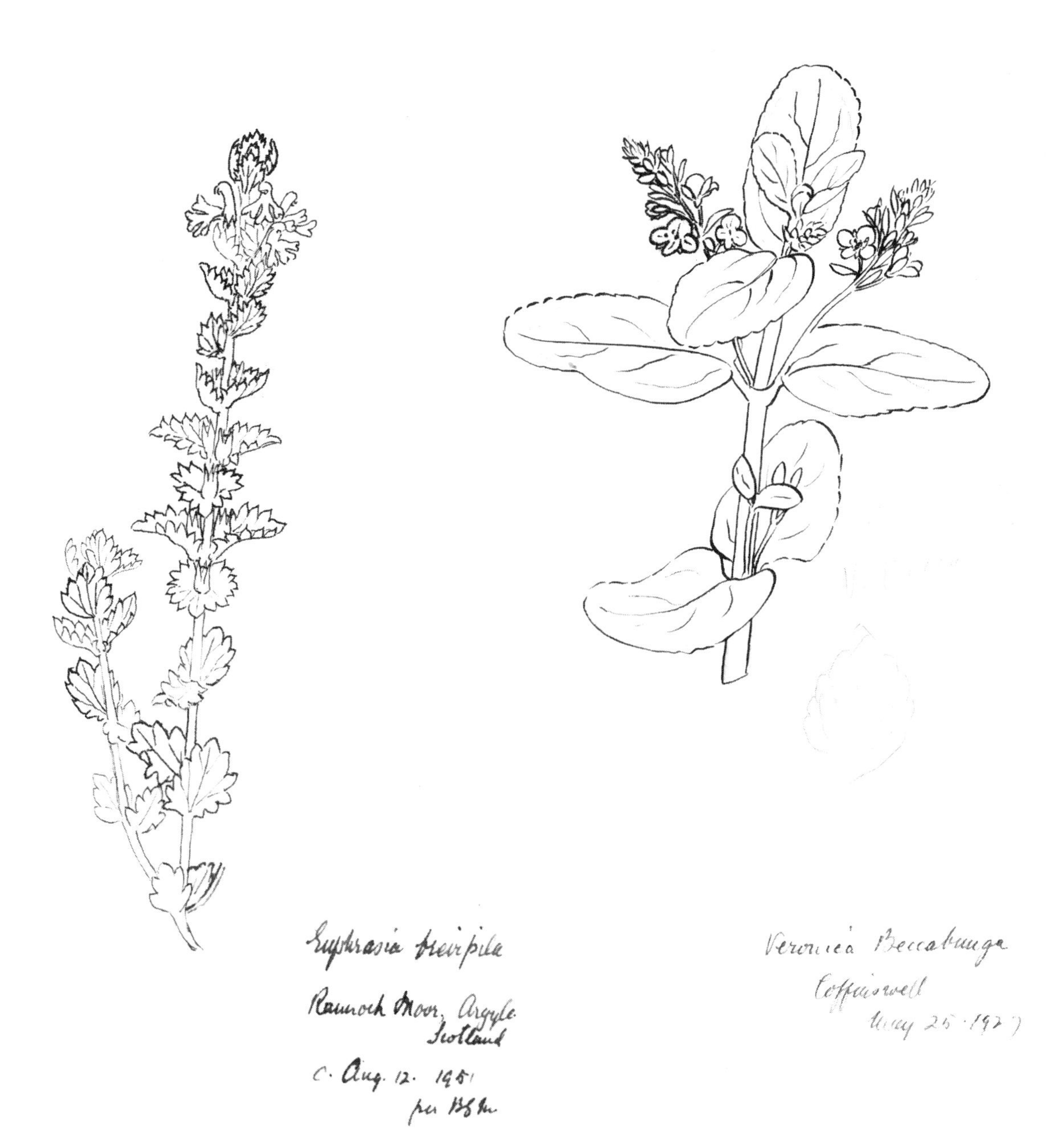

Euphrasia brevipila Burnat & Gremli.
Short-haired Eyebright

Veronica beccabunga L.
Brooklime

Melampyrum arvense L.
Field Cow-wheat

Pedicularis palustris L.
Red Rattle

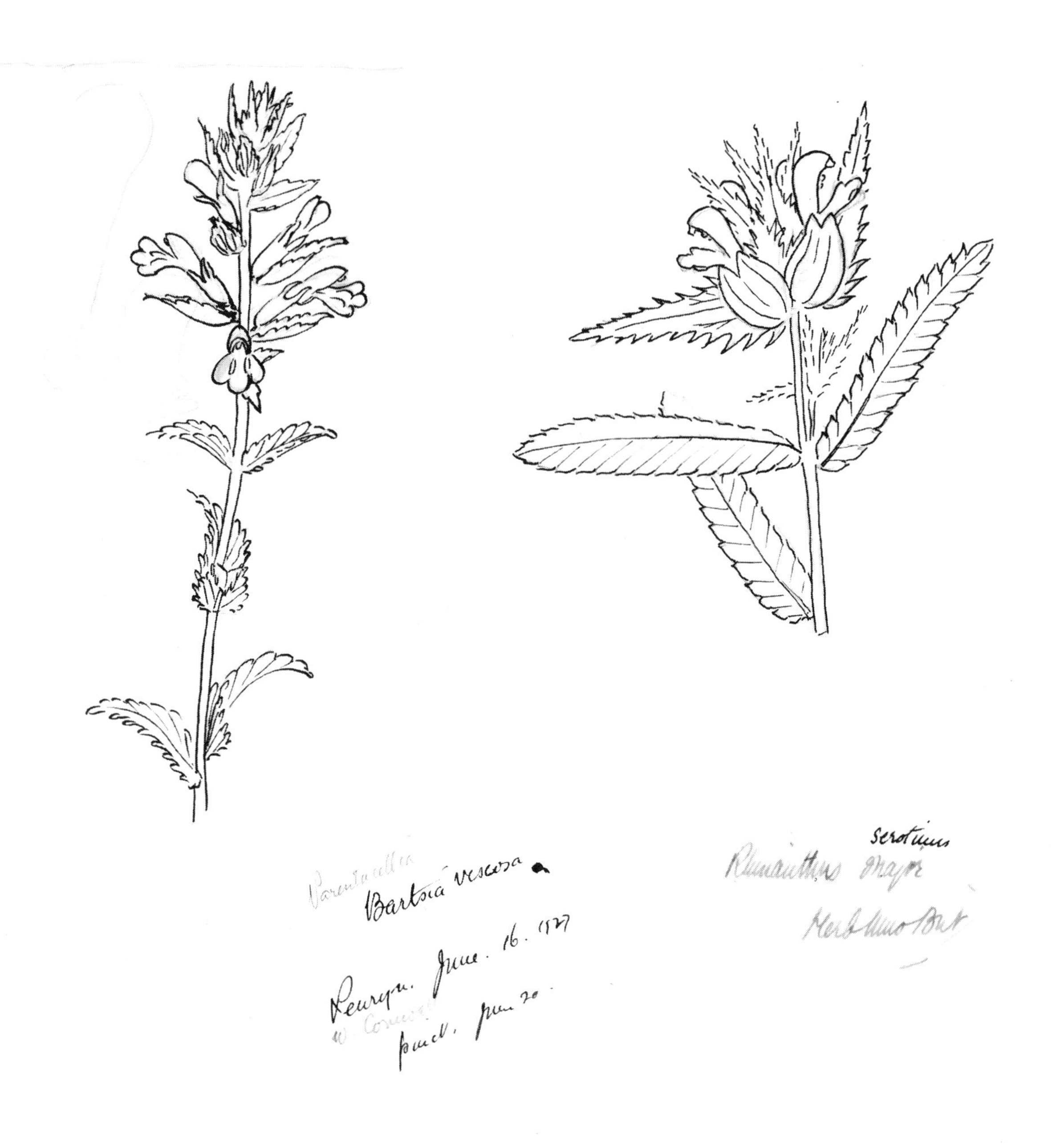

Bartsia viscosa L.
Yellow Bartsia

Rhinanthus serotinus (Schönh.) Oborny
Greater Hayrattle

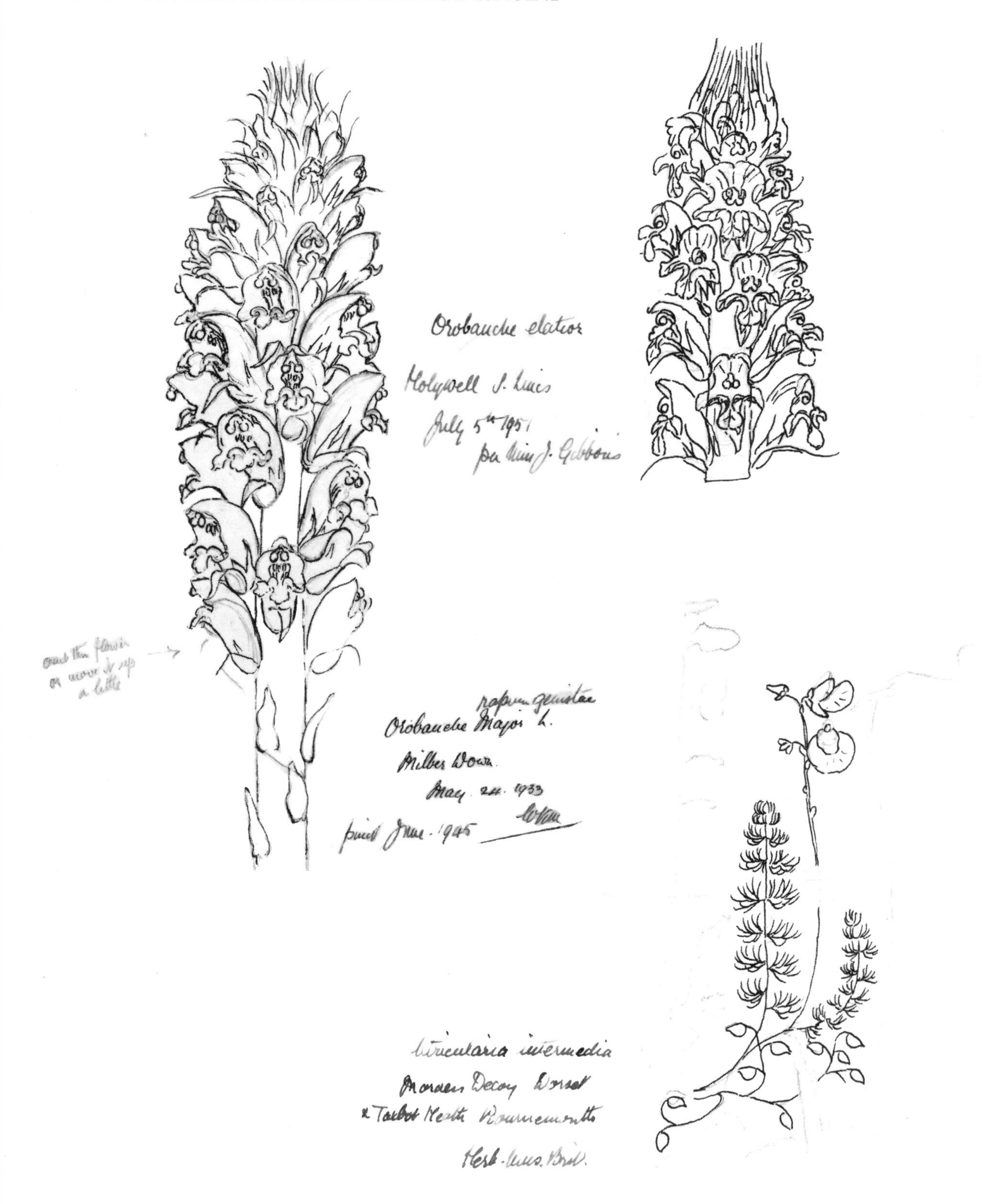

Orobanche rapum-genistae Thuill.
Greater Broomrape

Orobanche elatior Sutton.
Tall Broomrape

Utricularia intermedia Hayne.

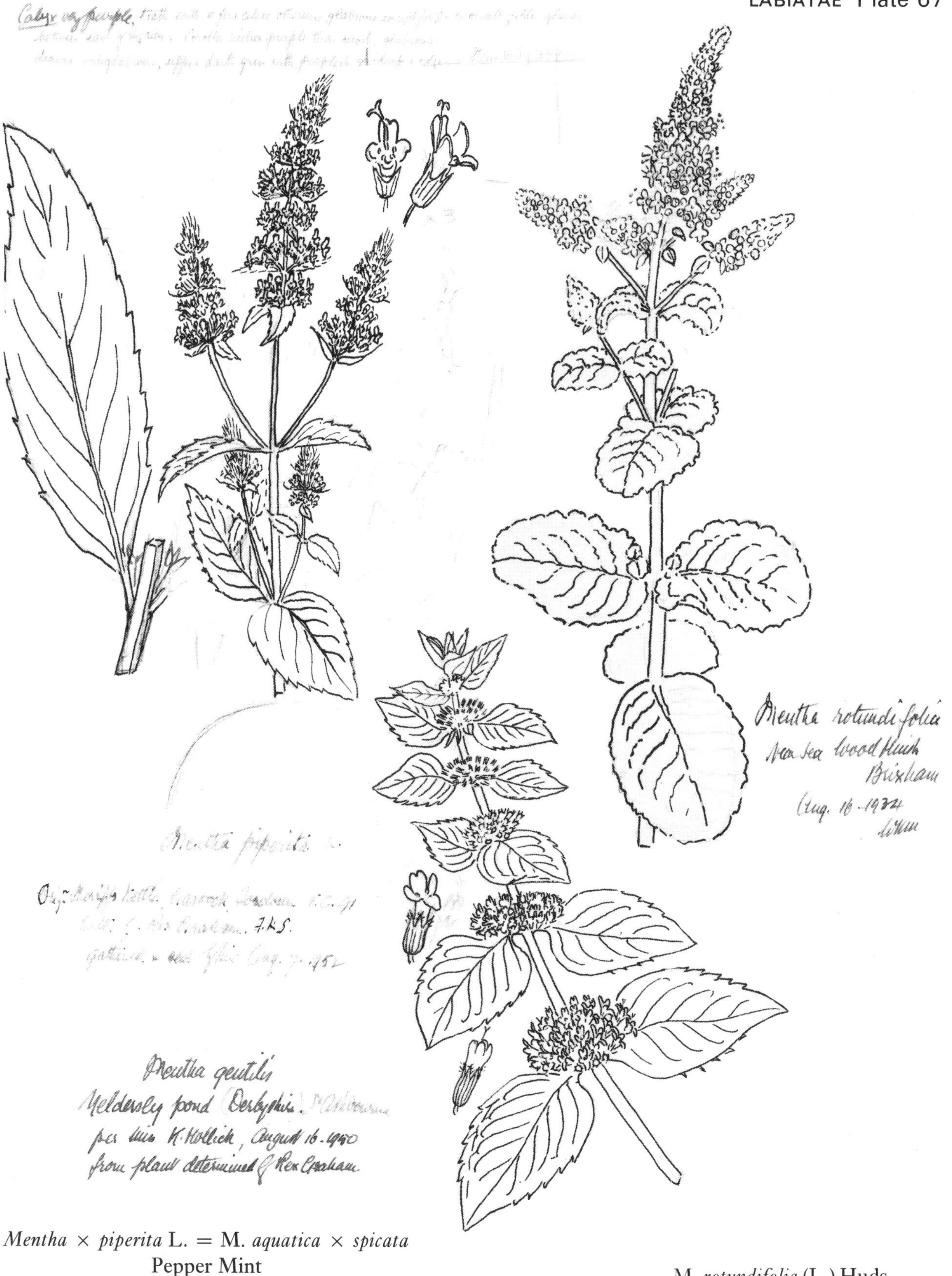

Mentha × *piperita* L. = M. *aquatica* × *spicata*
Pepper Mint

Mentha × *gentilis* L. = arvensis × spicata

M. *rotundifolia* (L.) Huds.
Round-leaved Mint.

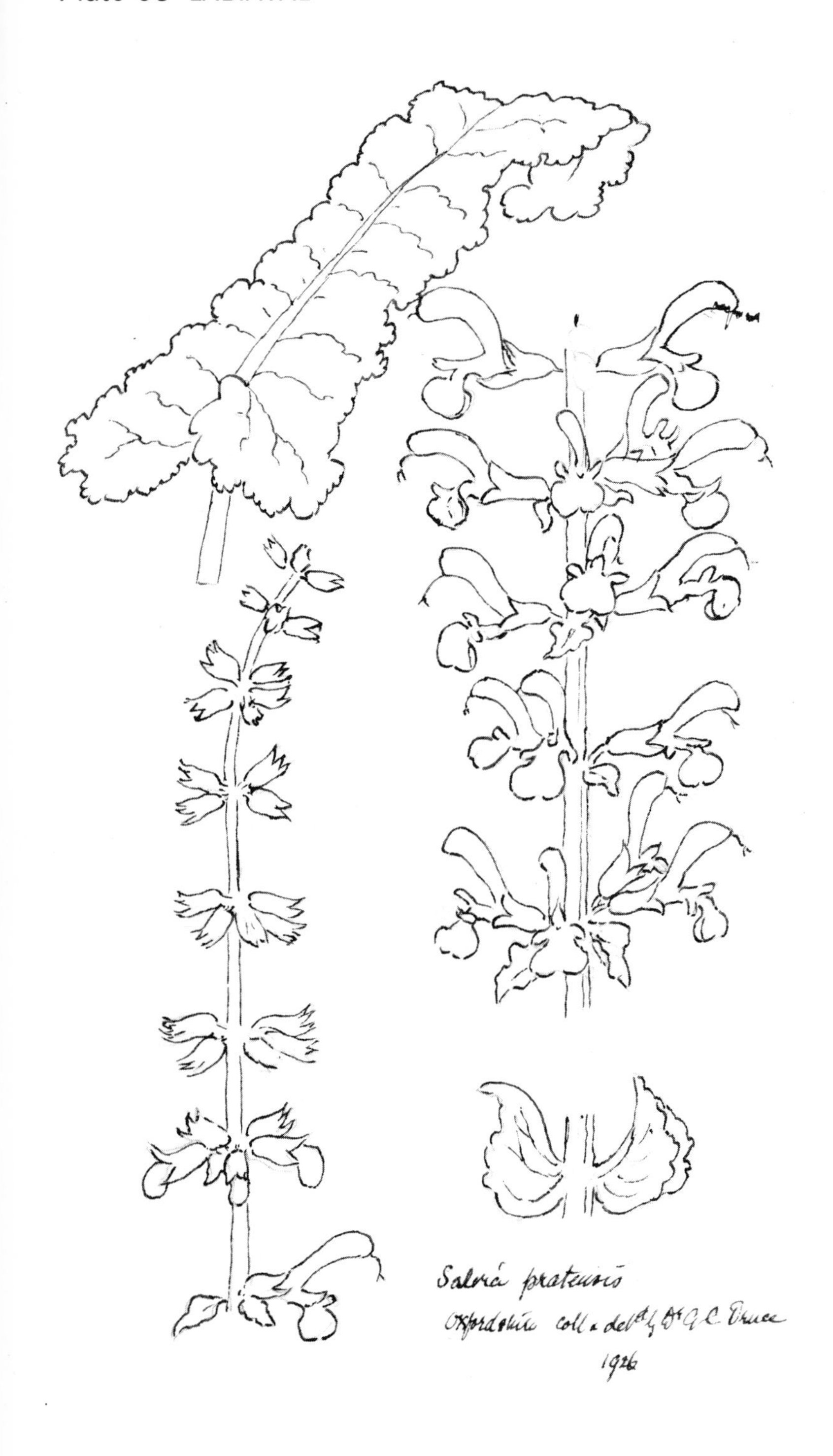

Salvia pratensis L.
Meadow Clary

Scutellaria galericulata L.
Skull-cap

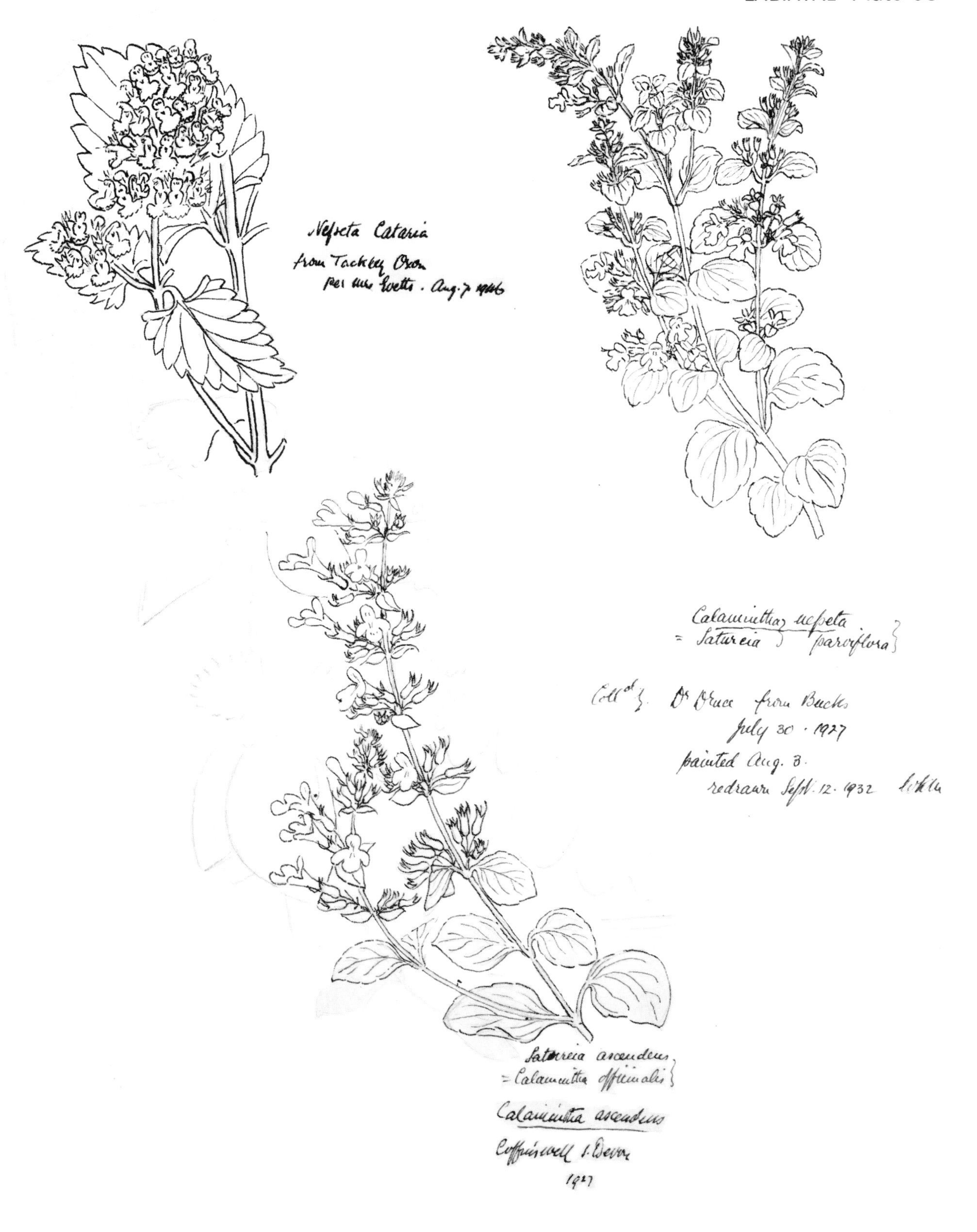

Nepeta cataria L.
Cat Mint

Calamintha ascendens Jord.
Common Calamint

Calamintha nepeta (L.) Savi.
Lesser Calamint

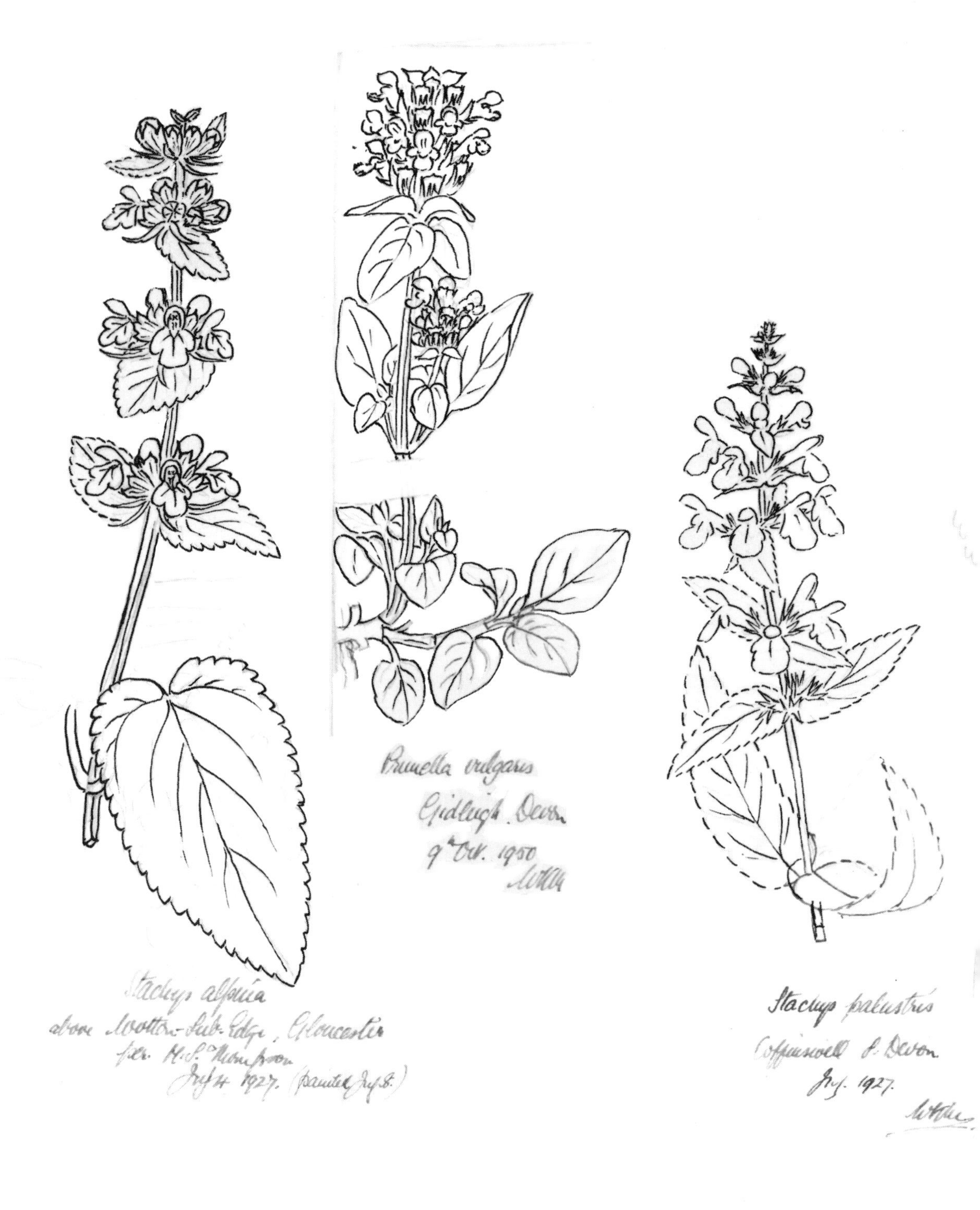

Stachys alpina L.
Alpine Woundwort

Prunella vulgaris L.
Self-heal

Stachys palustris L.
Marsh Woundwort

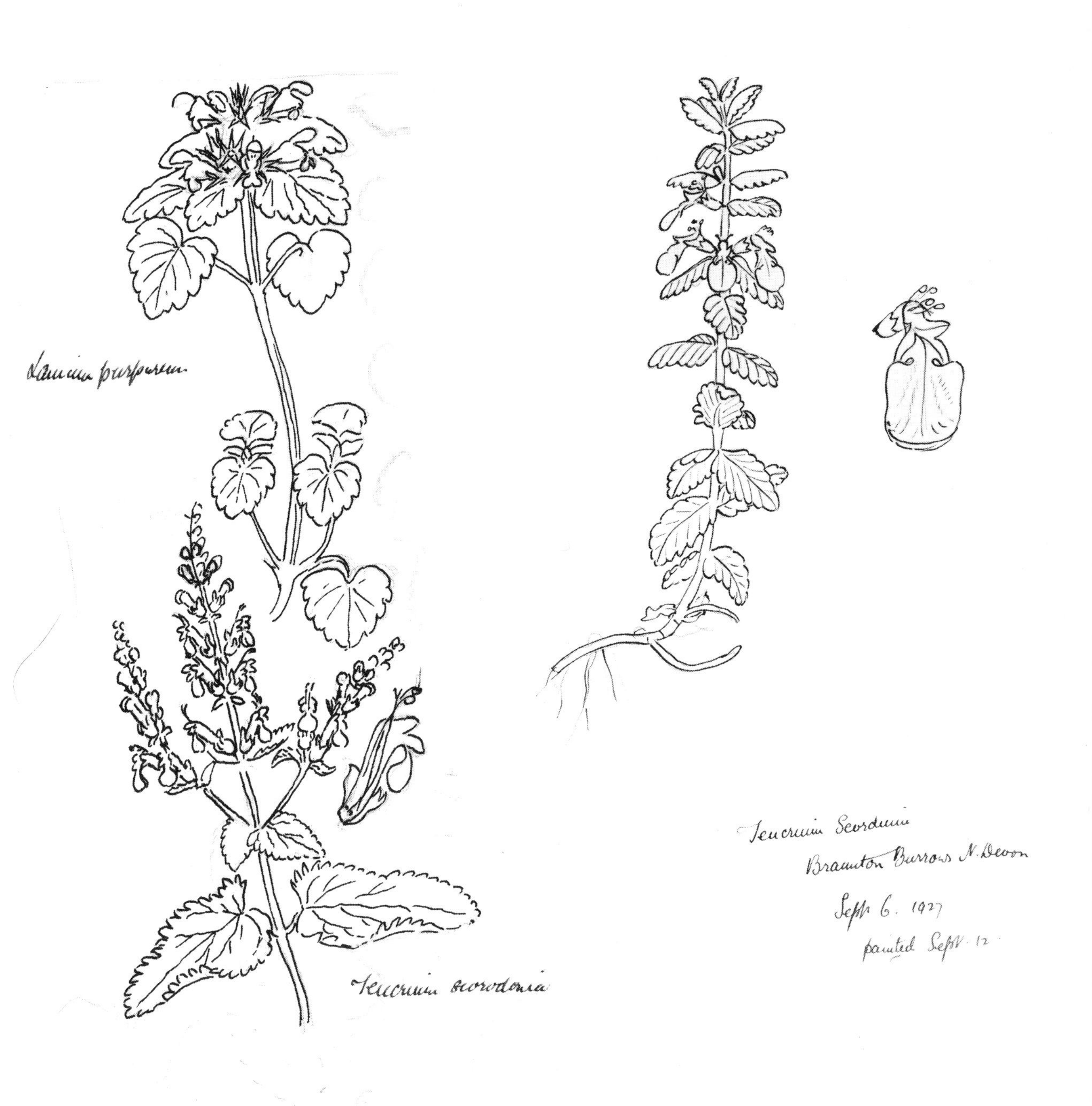

Lamium purpureum L.
Purple Dead-nettle

Teucrium scorodonia L.
Wood Sage

Teucrium scordium L.
Water Germander

Lamium amplexicaule

Melittis melissophyllum
Dartington S. Devon
June 24 1927

Lamium hybridum Vill.
Cut-leaved Dead-nettle

Melittis melissophyllum L.
Bastard Balm

Lamium amplexicaule L.
Henbit

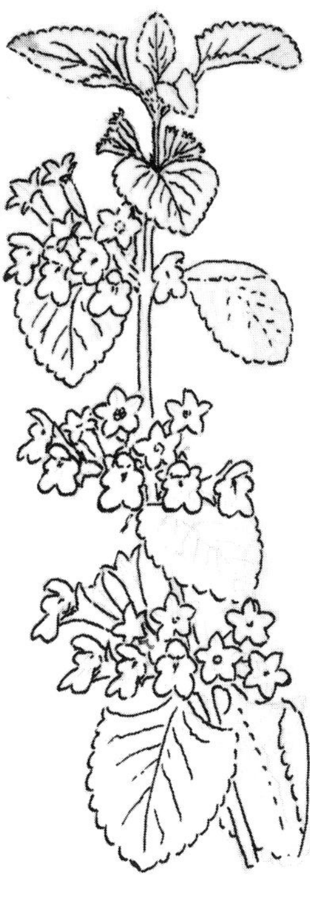

Lamium album L.
White Dead-nettle

Ajuga chamaepitys (L.) Schred.
Ground-pine

Ballota nigra L.
Horehound

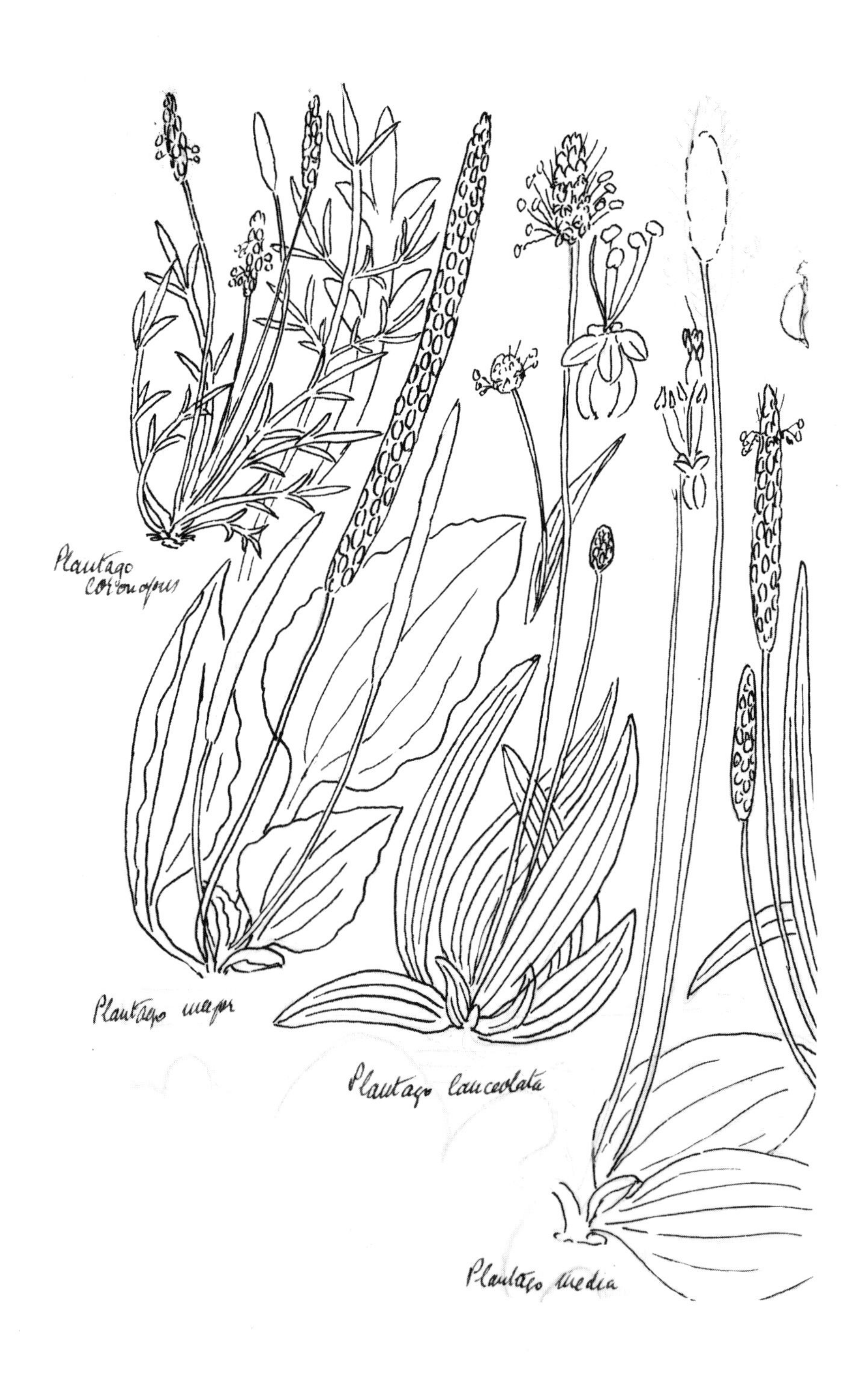

Redrawn from original sketches for the *Flora*

Chenopodium polyspermum L.	*Chenopodium album* L.	*Corrigiola litoralis* L.
Many-seeded Goose-foot	Fat Hen	Strap-wort

Chenopodium rubrum L.
Red Goose-foot

Chenopodium ficifolium Sm.
Fig-leaved Goose-foot

Atriplex patula L.
Common Orache

Atriplex hastata L.
Hastate Orache

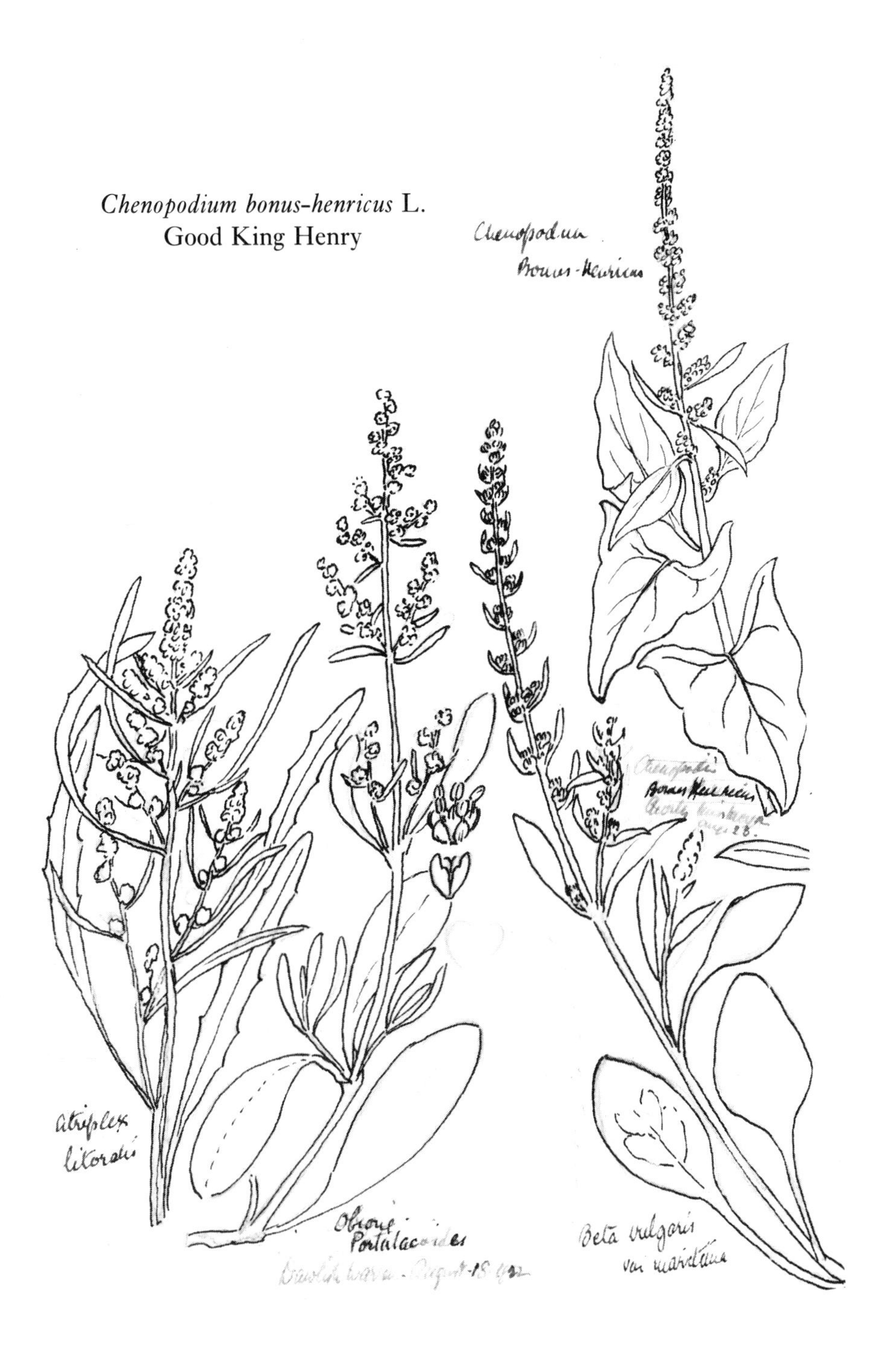

Chenopodium bonus-henricus L.
Good King Henry

Atriplex littoralis L.
Grass-leaved Orache

Halimione portulacoides (L.) Aellen
Sea Purslane

Beta maritima L.
Sea Beet

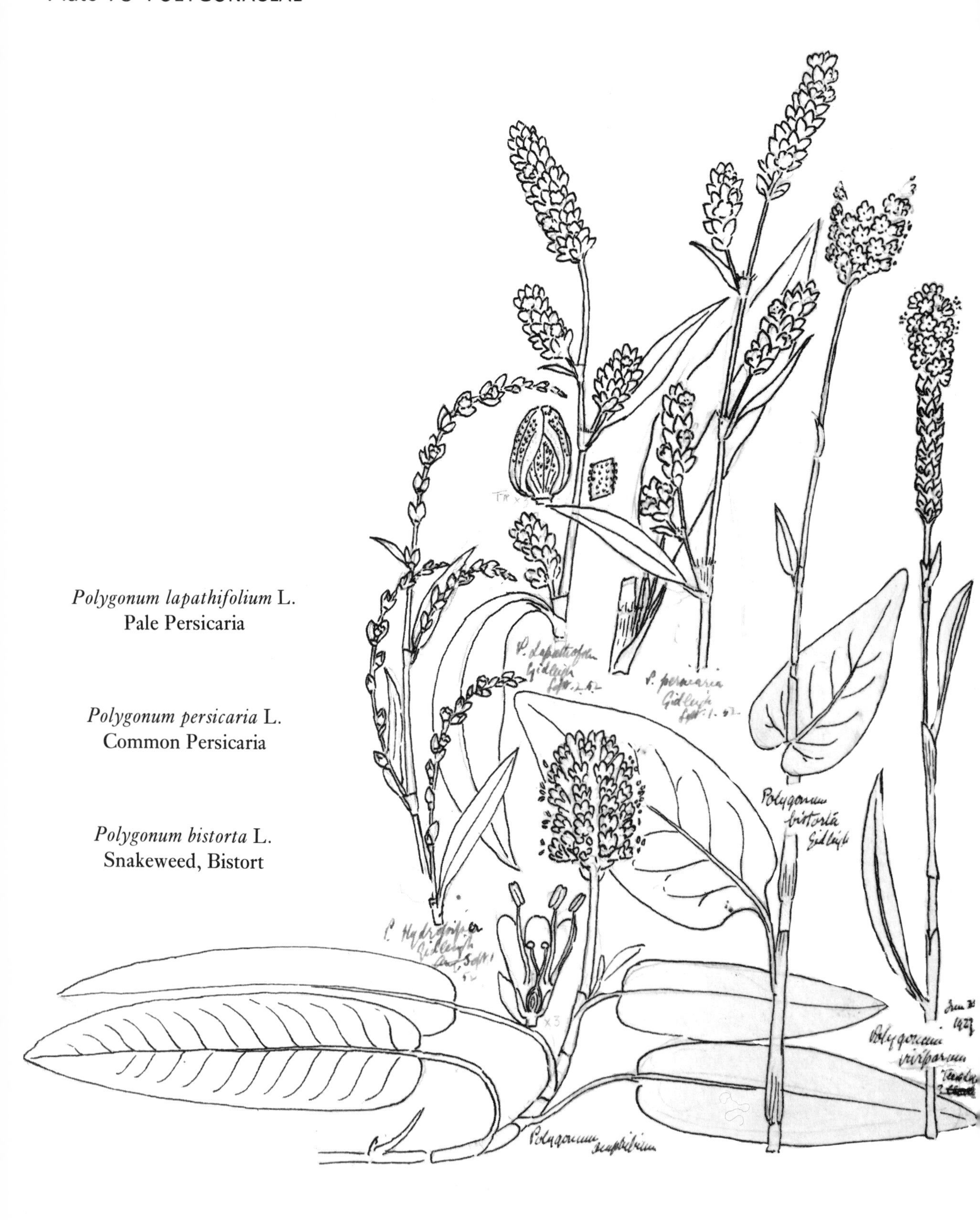

Polygonum lapathifolium L.
Pale Persicaria

Polygonum persicaria L.
Common Persicaria

Polygonum bistorta L.
Snakeweed, Bistort

Polygonum hydropiper L.
Water Pepper, Biting Persicaria

Polygonum amphibium L.
Amphibious Persicaria

Polygonum viviparum L.
Alpine Bistort

Polygonum nodosum Pers.
Spotted Persicaria

Redrawn from original sketches for the *Flora*

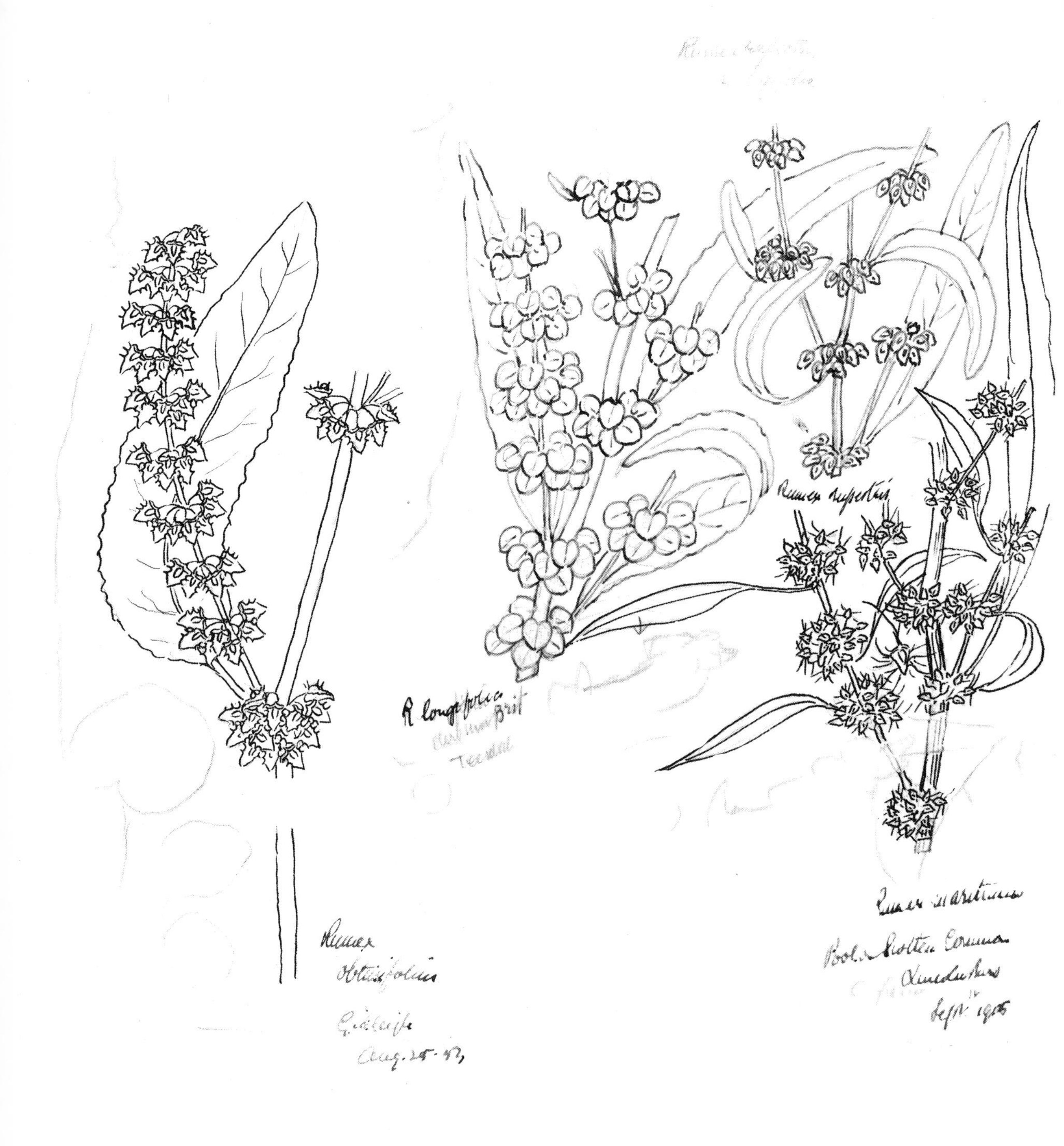

Rumex obtusifolius L.
Broad-leaved Dock

Rumex longifolius DC.
Long-leaved Dock

Rumex rupestris Le Gall.
Shore Dock

Rumex maritimum L.
Golden Dock

Rumex conglomeratus Murr.
Sharp Dock

Rumex acetosa L.
Common Sorrel

Rumex acetosella L.
Sheep's Sorrel

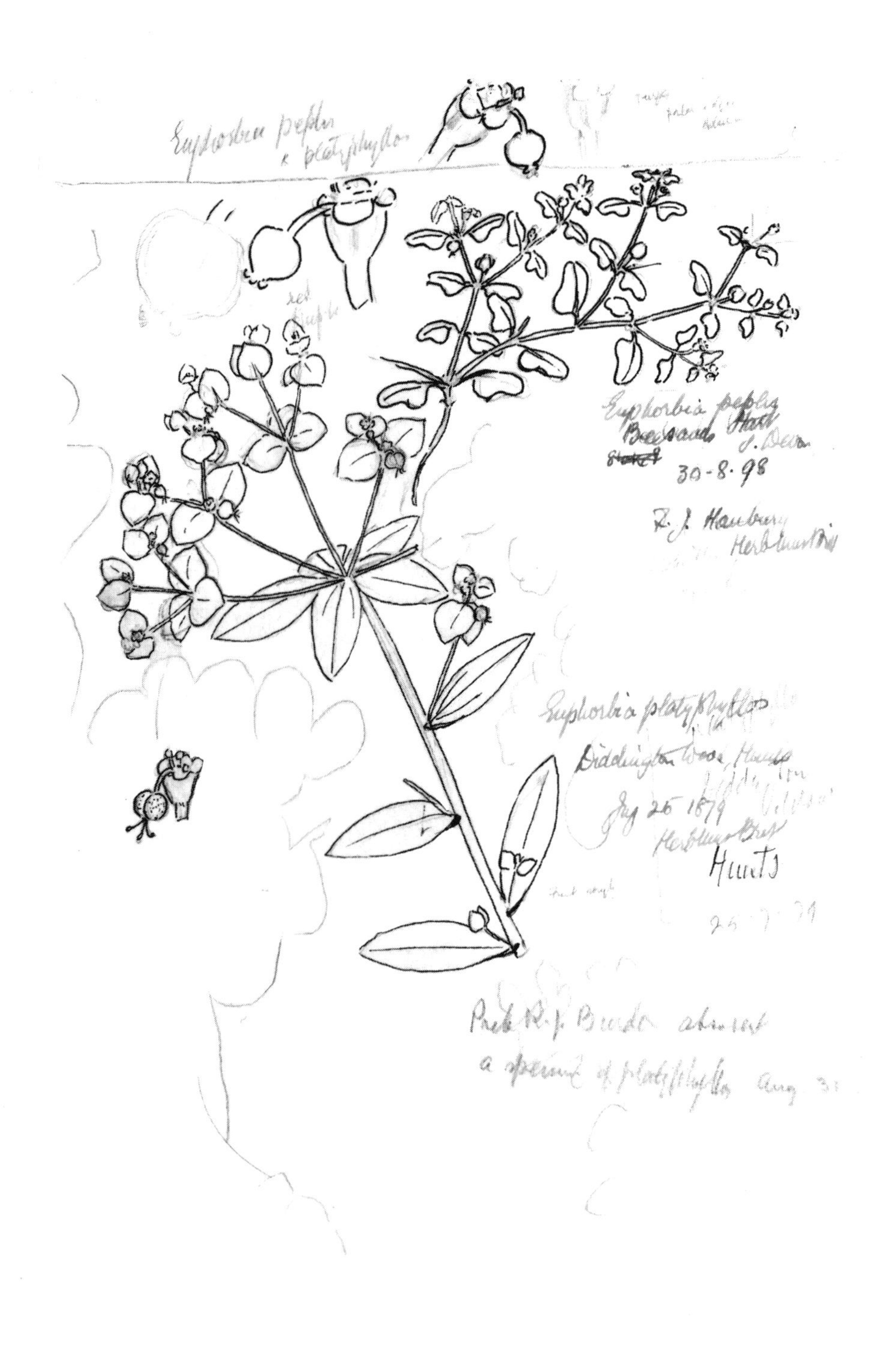

Euphorbia peplis L.
Purple Spurge

Euphorbia platyphyllos L.
Broad-leaved Spurge

Euphorbia hyberna L.
Irish Spurge

Euphorbia helioscopia L.
Sun Spurge

Euphorbia paralias L.
Sea Spurge

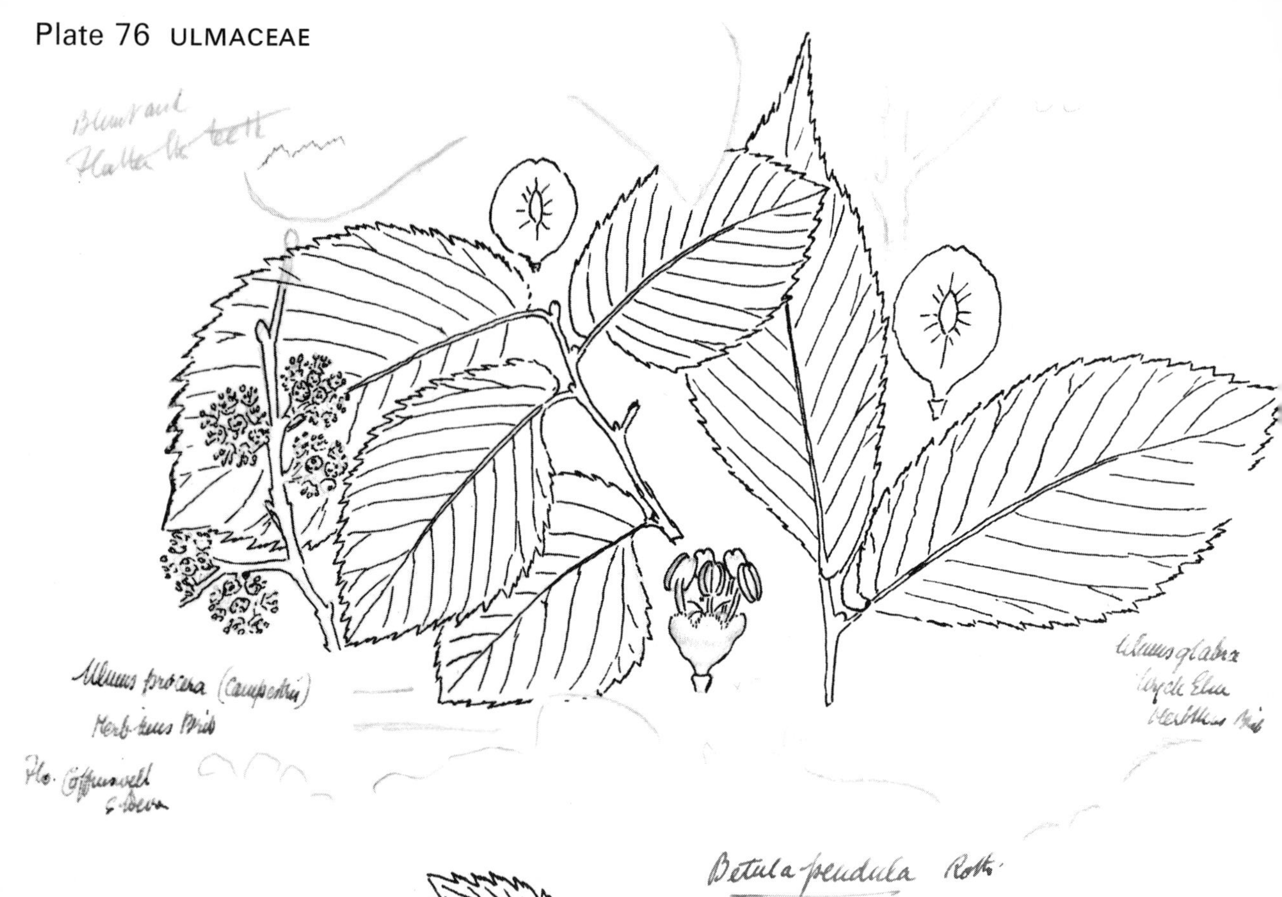

Ulmus procera Salisb.
English Elm

Betula pendula Roth.
Silver Birch

Ulmus glabra Huds.
Wych Elm

Urtica dioica L.
Stinging Nettle

Parietaria diffusa Mert. & Kock.
Pellitory of the Wall

Betula pubescens Ehrh.
Brown Birch

Plate 77 BETULACEAE, FAGACEAE and CORYLACEAE

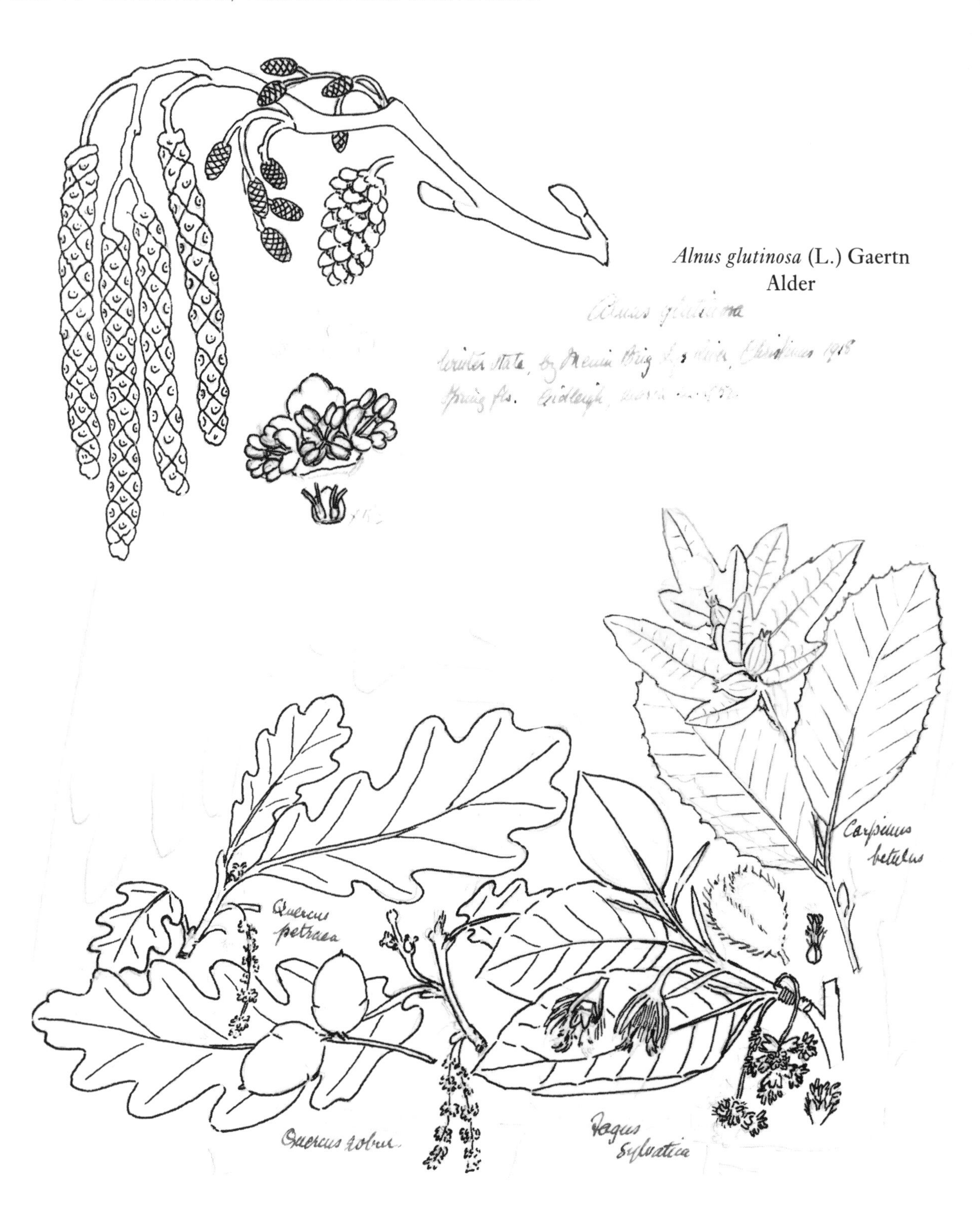

Redrawn from original sketches for the *Flora*

Salix lapponum L.
Downy Willow

Salix trianda L.
Almond Willow

Salix fragilis L.
Crack Willow

Salix arbuscula
Ben Lawers, Perth. July 9th 1953
per A. W. Stelfox

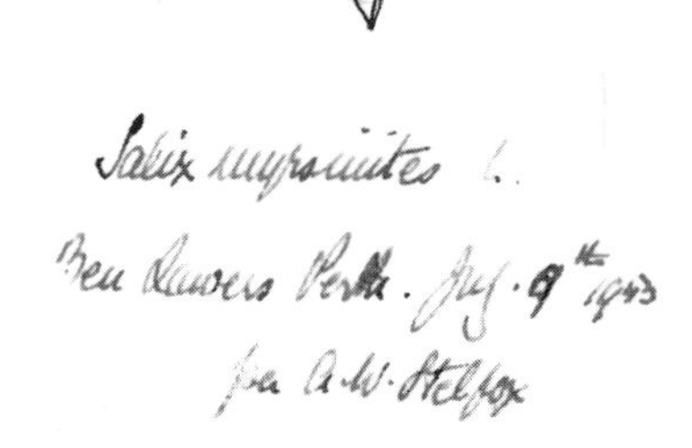

Salix nigricans. Ben Lawers per A.W. Stelfox
July. 19. 1963

Salix arbuscula L.
Little Tree Willow

Salix nigricans Sm.
Dark-leaved Willow

Salix myrsinites L.
Myrtle-leaved Willow

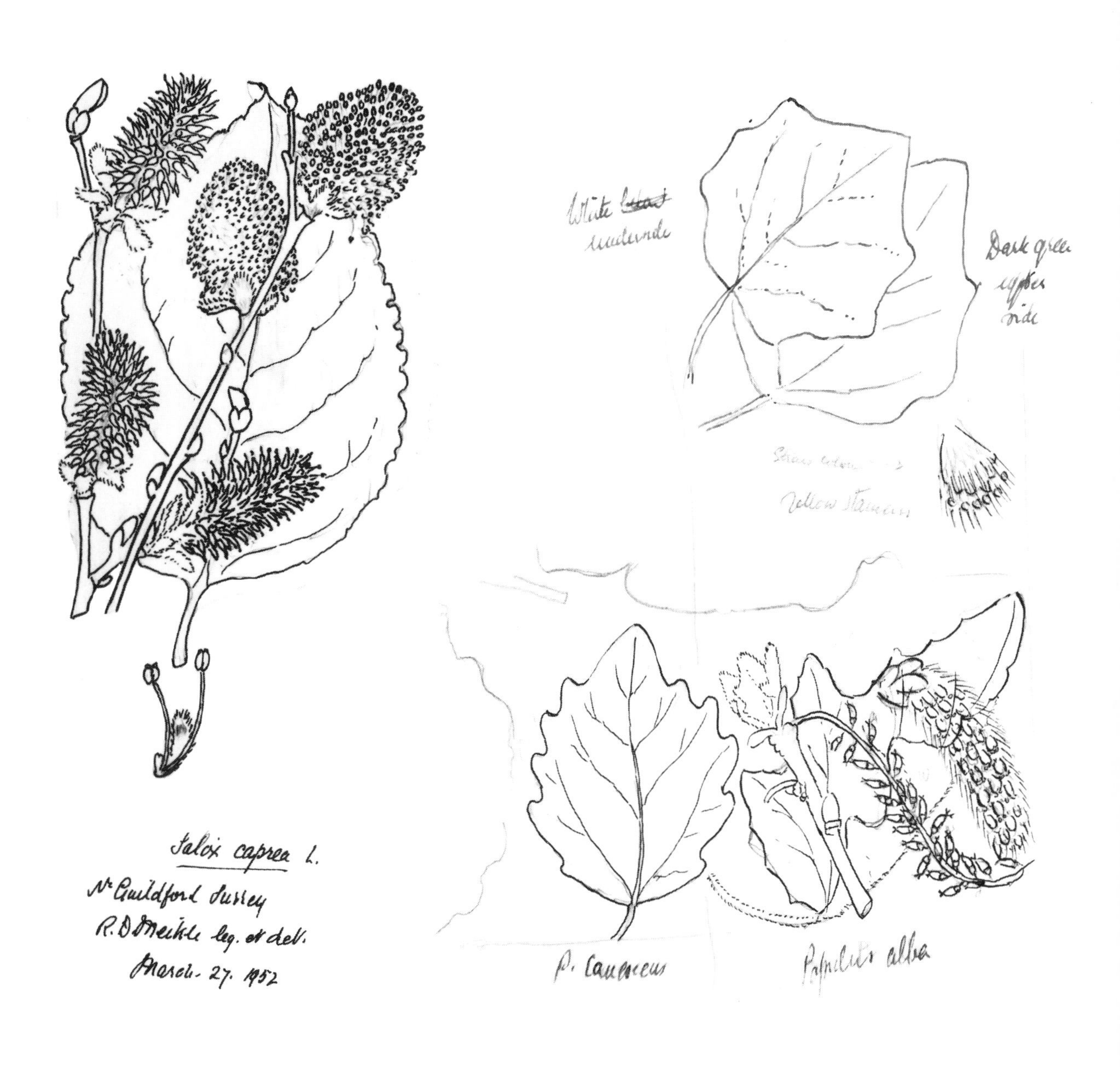

Salix caprea L.
Great Sallow

Populus canescens (Ait.) Sm.
Grey Poplar

Populus alba L.
Lobed or White Poplar

Plate 79 BUTOMACEAE and ALISMATACEAE

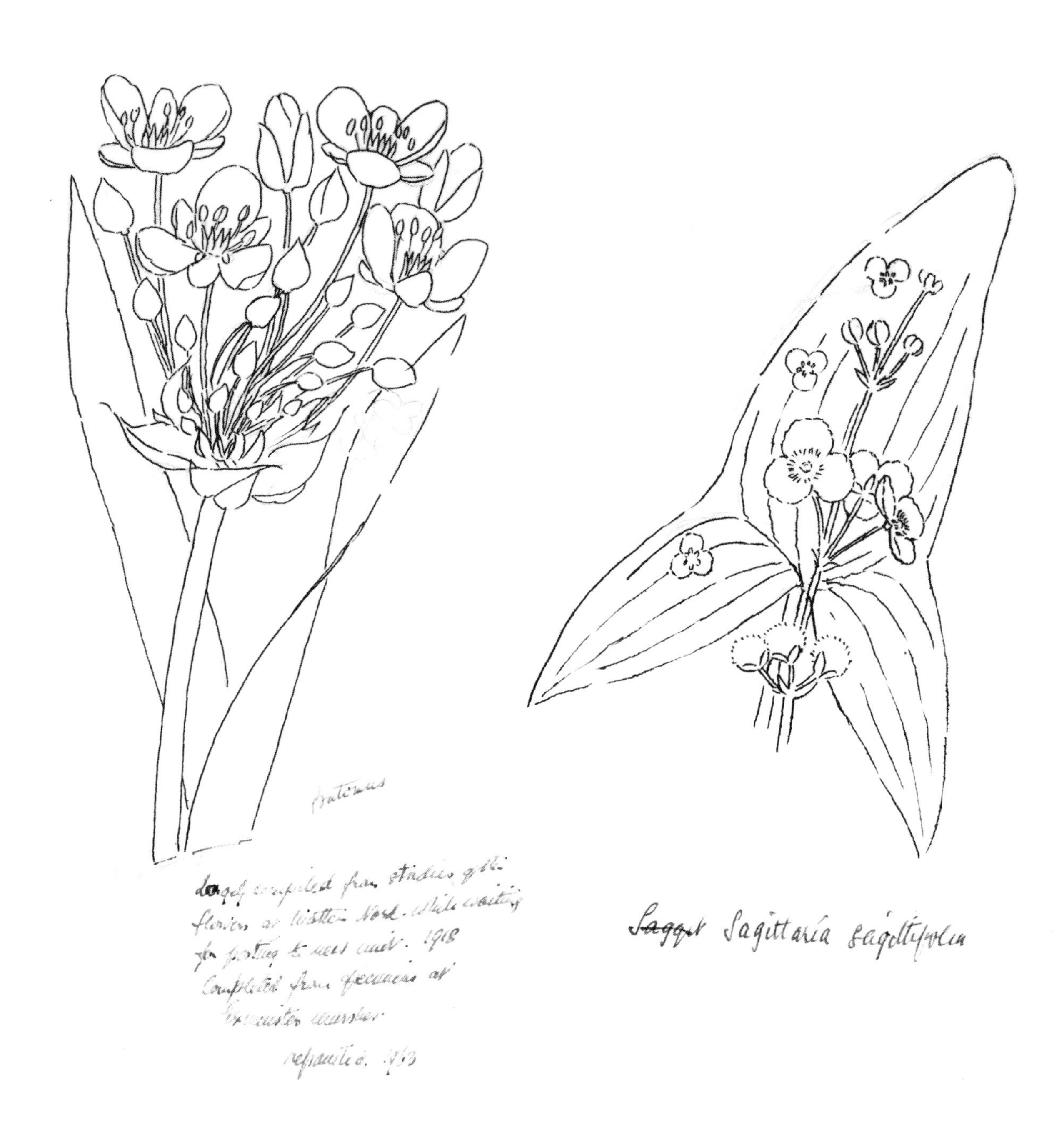

Butomus umbellatus L.
Flowering Rush

Sagittaria sagittifolia L.
Arrowhead

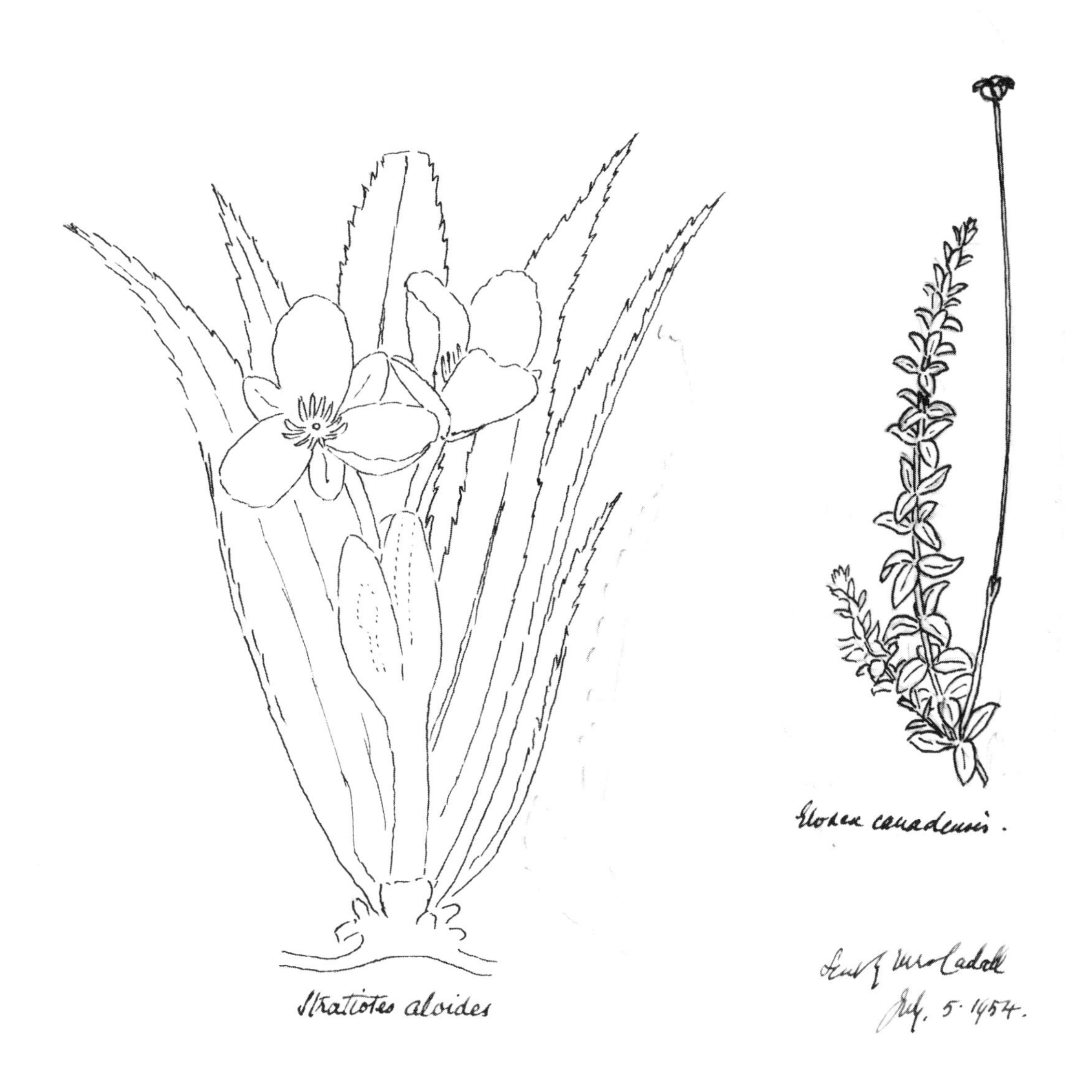

Stratiotes aloides L.
Water Soldier

Elodea canadensis Michx.
Canadian Pondweed

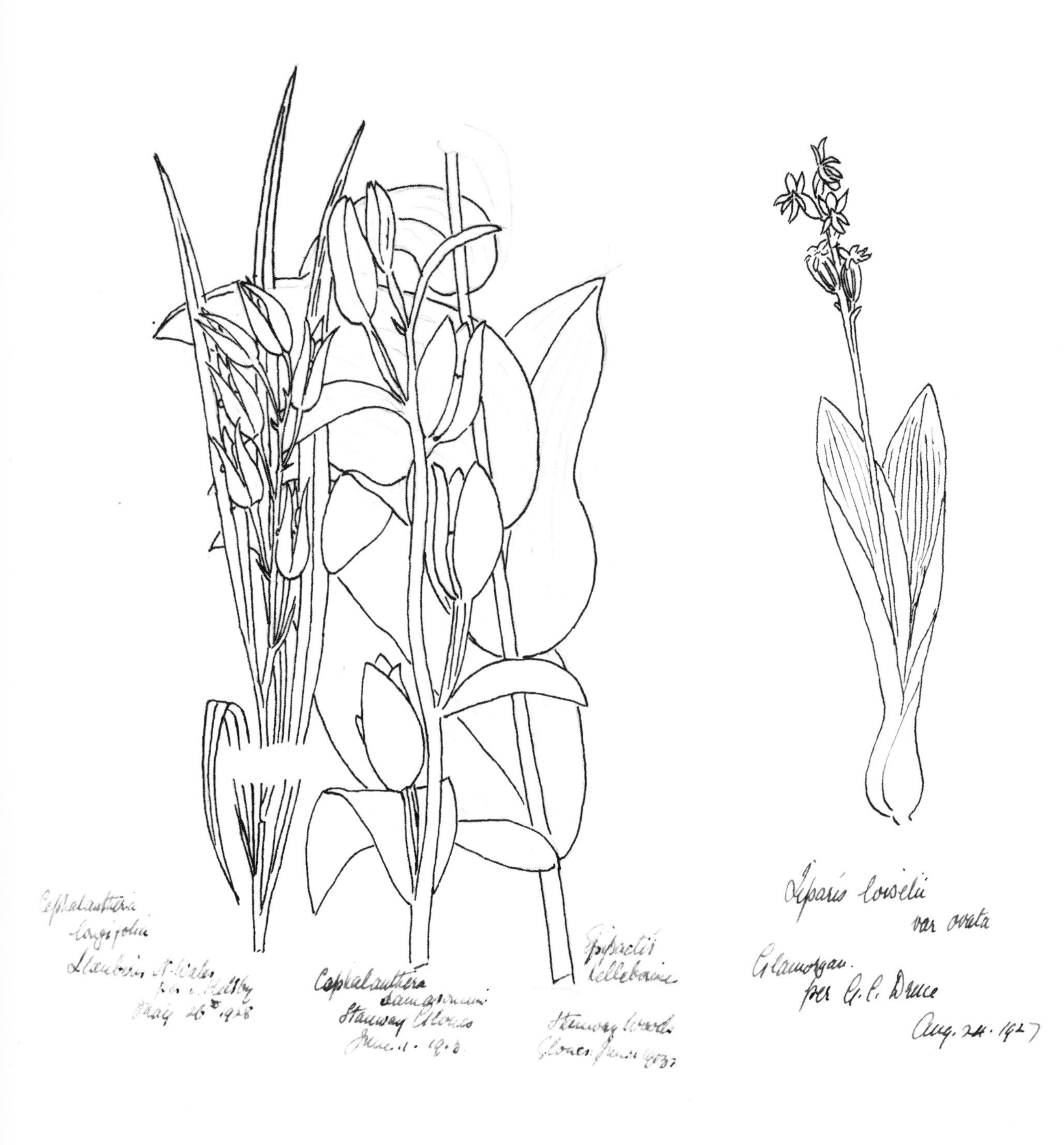

Cephalanthera longifolia (L.) Fritsch
Long-leaved Helleborine

Cephalanthera damasonium (Mill.) Druce
White Helleborine

Liparis loeselii (L.) Rich.
Fen Orchid

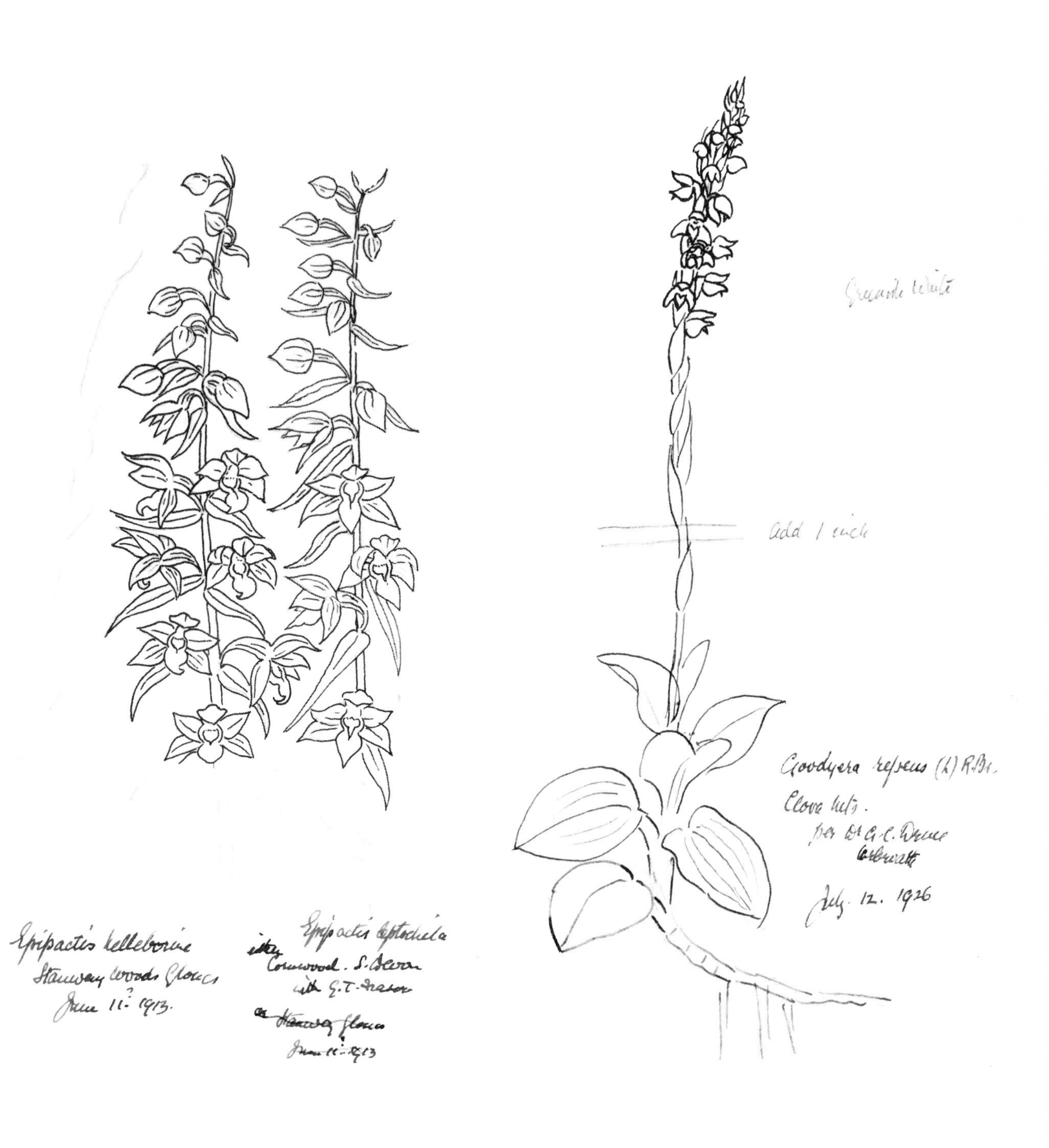

Epipactis helleborine (L.) Crantz.
Broad Helleborine

Epipactis leptochila (Godfery) Godfery
Narrow-lipped Helleborine

Goodyera repens (L.) R. Br.
Creeping Lady's Tresses

Epipactis palustris (L.) Crantz.
Marsh Helleborine

Dactylorhiza incarnata (L.) Soó

Redrawn from original sketches for the *Flora*

Himantoglossum hircinum (L.) Spreng.
Lizard Orchid

Aceras anthropophorum L.	*Platanthera bifolia* (L.) Rich.	*Leucorchis albida* (L.) E. Mey. ex Schur.
Man Orchid	Lesser Butterfly Orchid	Small White Orchid

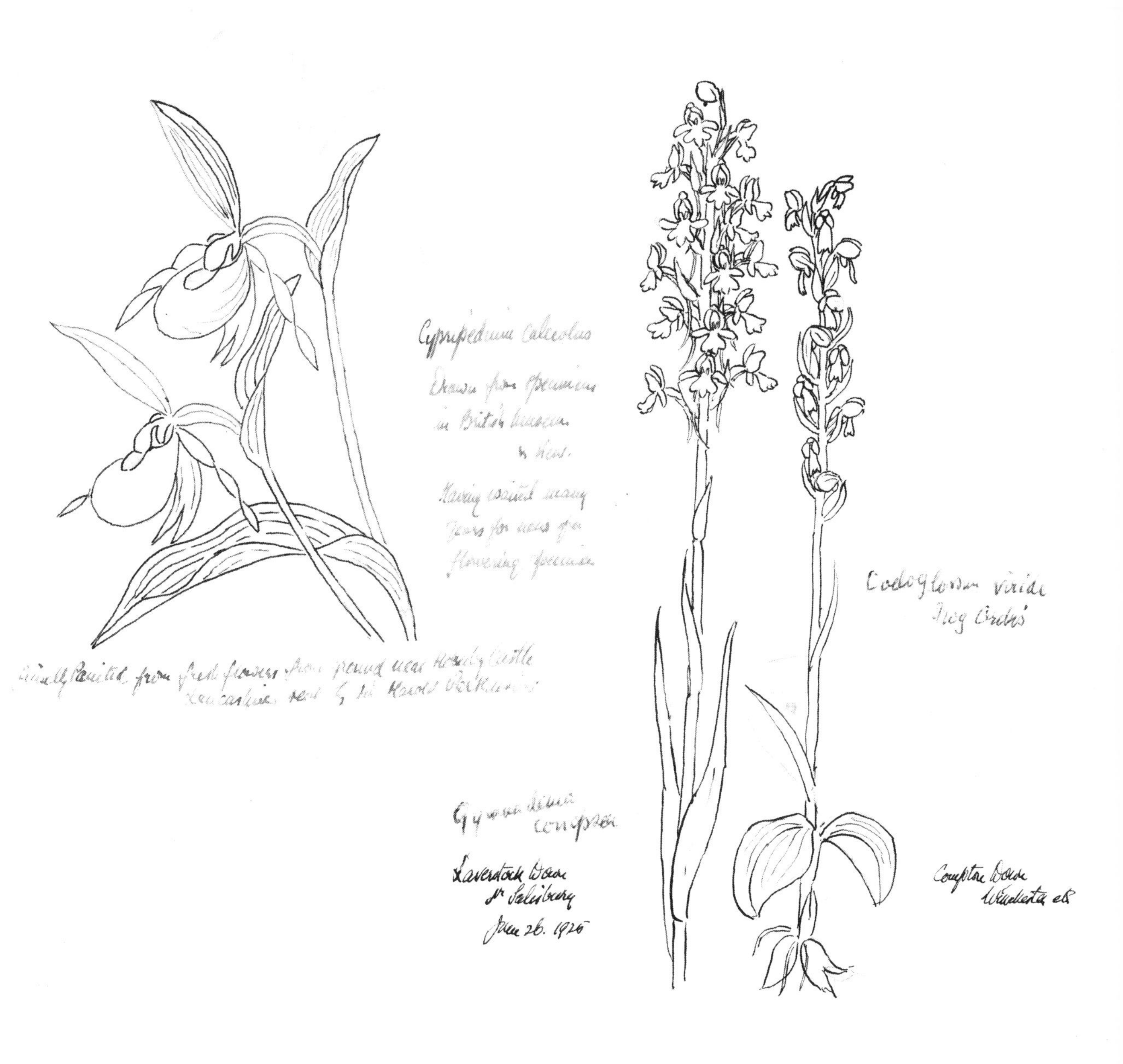

Cypripedium calceolus L. Lady's Slipper	*Gymnadenia conopsea* (L.) R. Br. Sweet-scented Orchid	*Coeloglossum viride* (L.) Hartm. Frog Orchid

Plate 83 AMARYLLIDACEAE and IRIDACEAE

Redrawn from original sketches for the *Flora*

Iris foetidissima L.
Stinking Iris

Romulea columnae
Dawlish Warren S. Devon
April 24 1925
& successive years

Copy in Flora of Devon

Romulea columnae Seb & Mauri
Warren Crocus

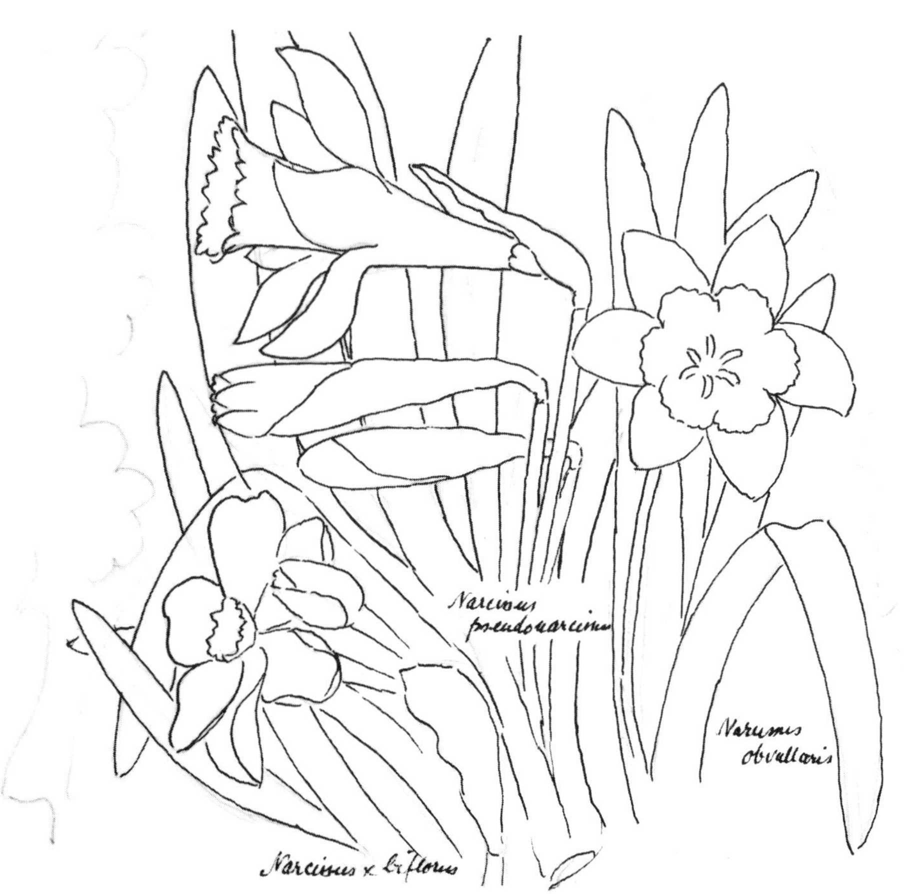

Redrawn from original sketches for the *Flora*

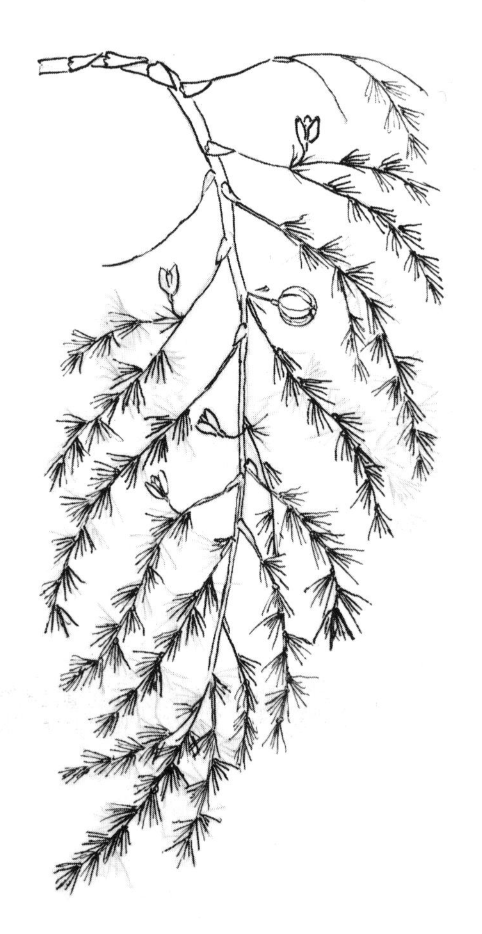

Asparagus

Lizard Cornwall

June 16. 1928

Ruscus aculeatus.

S. Devon 1933

Asparagus officinalis L.
Subsp. *prostratus* (Dumort.) E. F. Warb.
Wild Asparagus

Ruscus aculeatus L.
Butcher's Broom

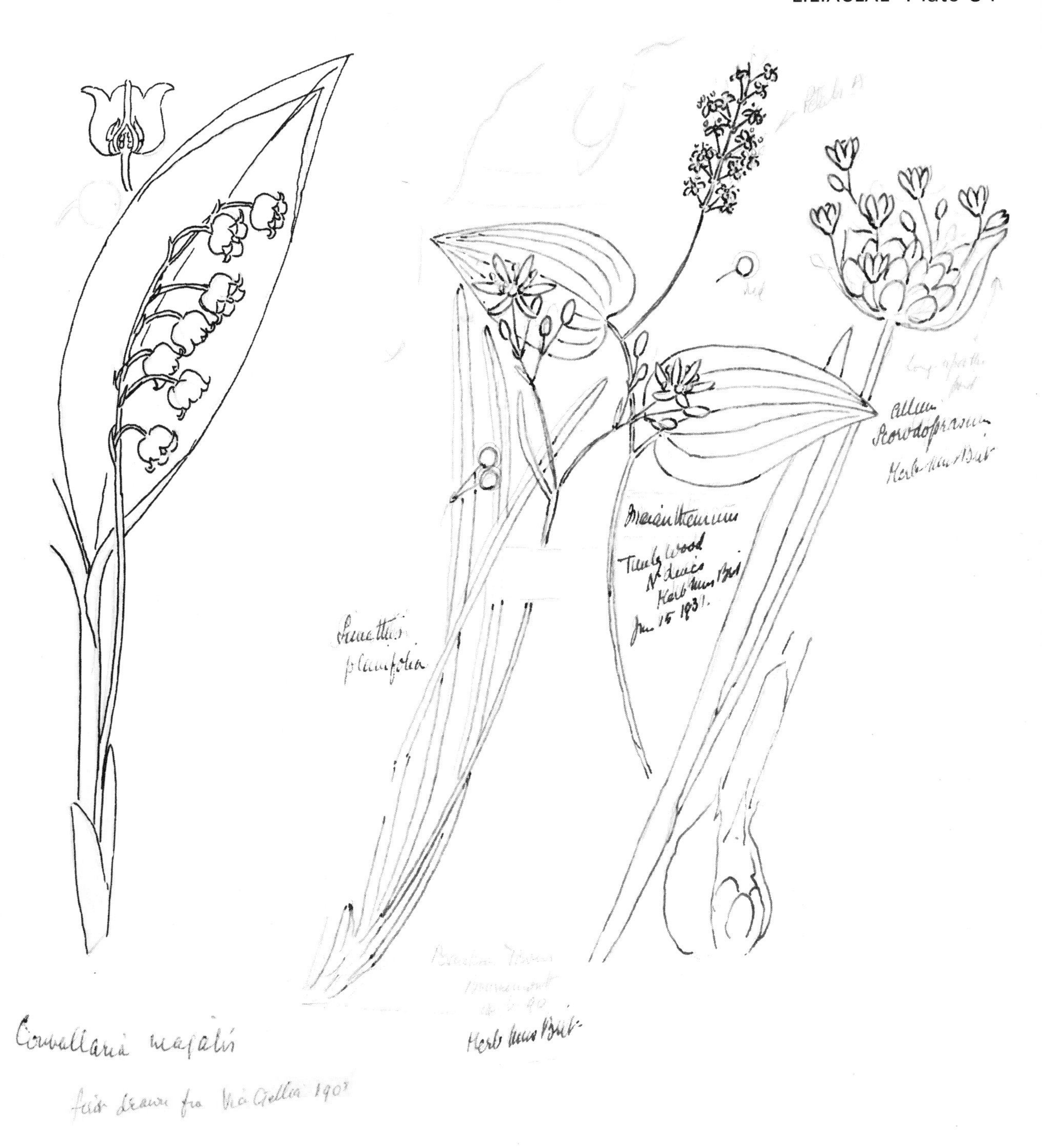

Simethis planifolia (L.) Gren. & Godr.

Convallaria majalis L.
Lily of the Valley

Maianthemum bifolium (L.) Schmidt.
May Lily

Allium scorodoprasum L.
Sand Leek

Lilium pyrenaicum Gouan
Pyrenean Lily

Ornithogalum nutans L.
Drooping Star of Bethlehem

Allium triquetrum L.
Triangular-stalked Garlic

Tulipa sylvestris L.
Wild Tulip

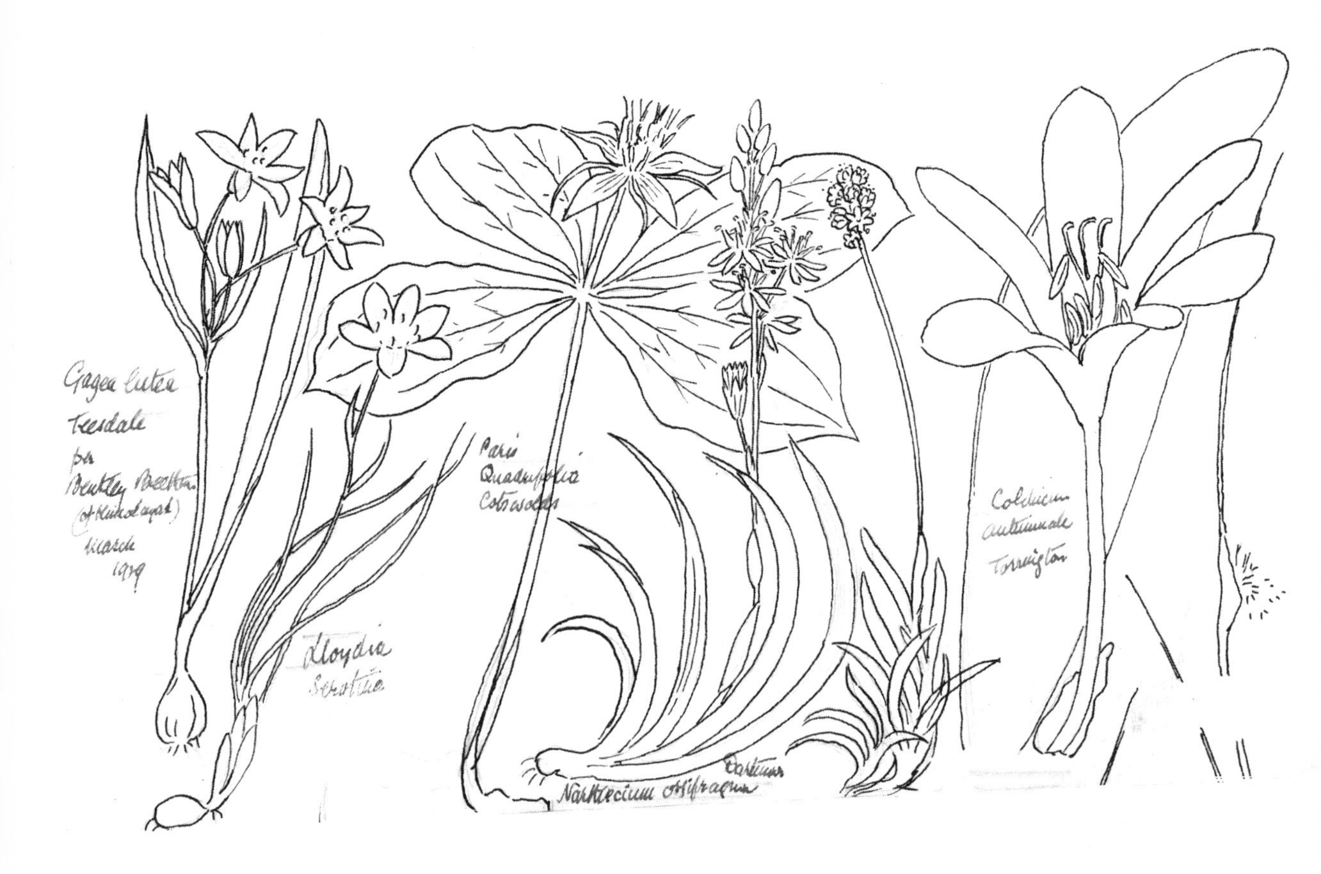

Redrawn from original sketches for the *Flora*

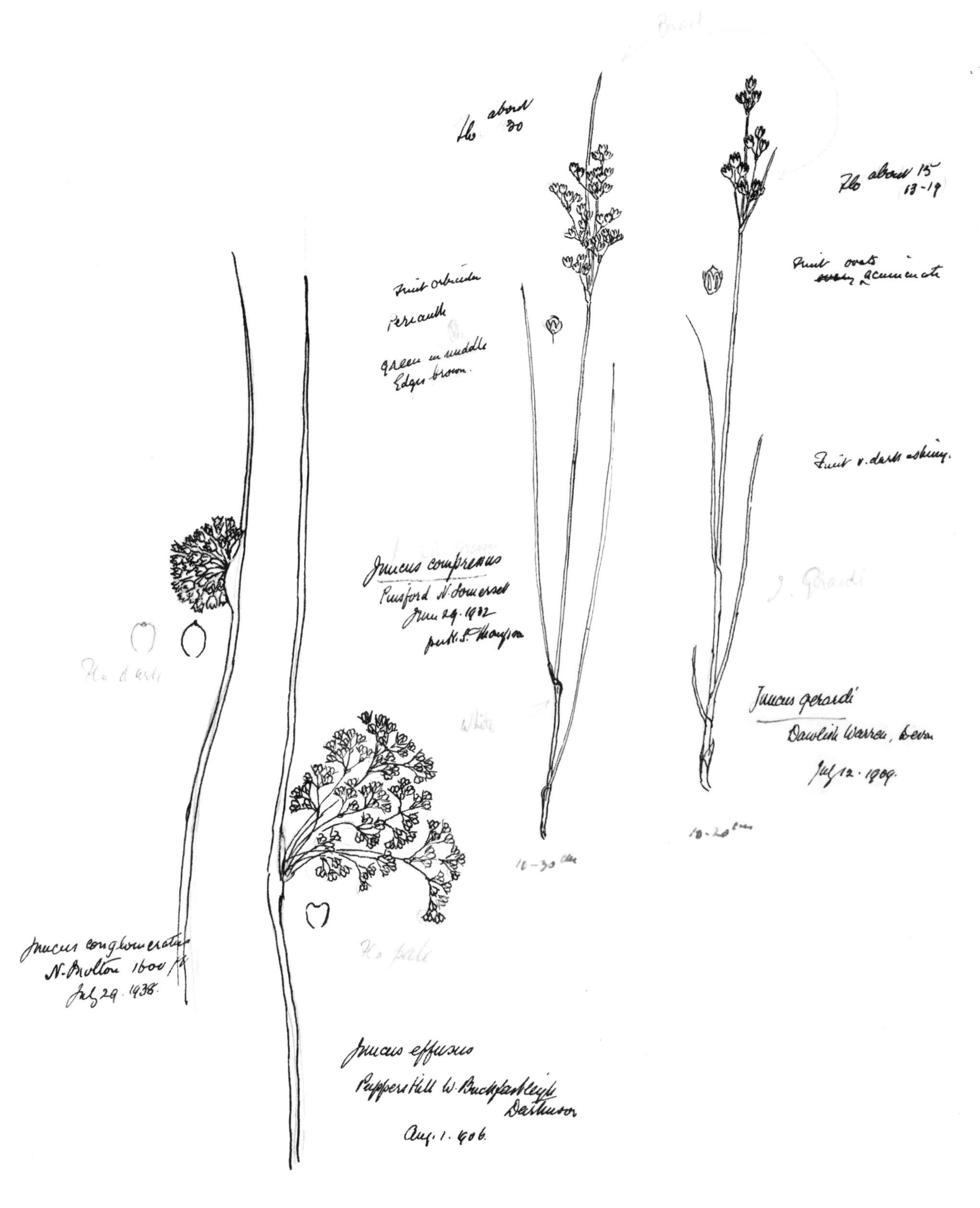

Juncus conglomeratus L.
Common Rush

Juncus effusus L.
Soft Rush

Juncus compressus Jaq.
Round-fruited Rush

Juncus gerardii Lois.
Salt Mud Rush

Juncus maritimus Lam.
Sea Rush

Juncus acutus L.
Sharp Sea Rush

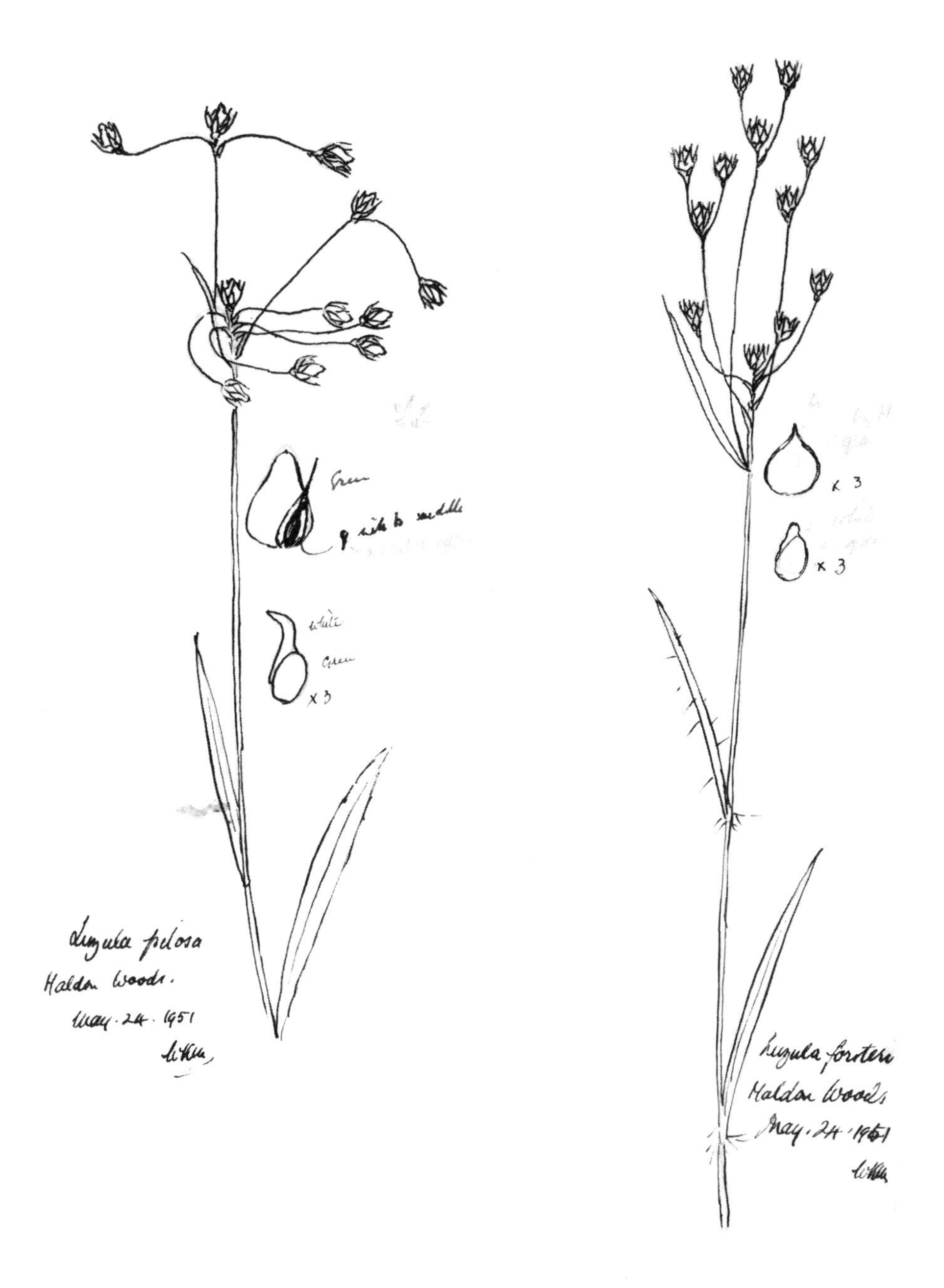

Luzula pilosa (L.) Wild.
Hairy Wood-rush

Luzula forsteri (Sm.) DC.
Forster's Wood-rush

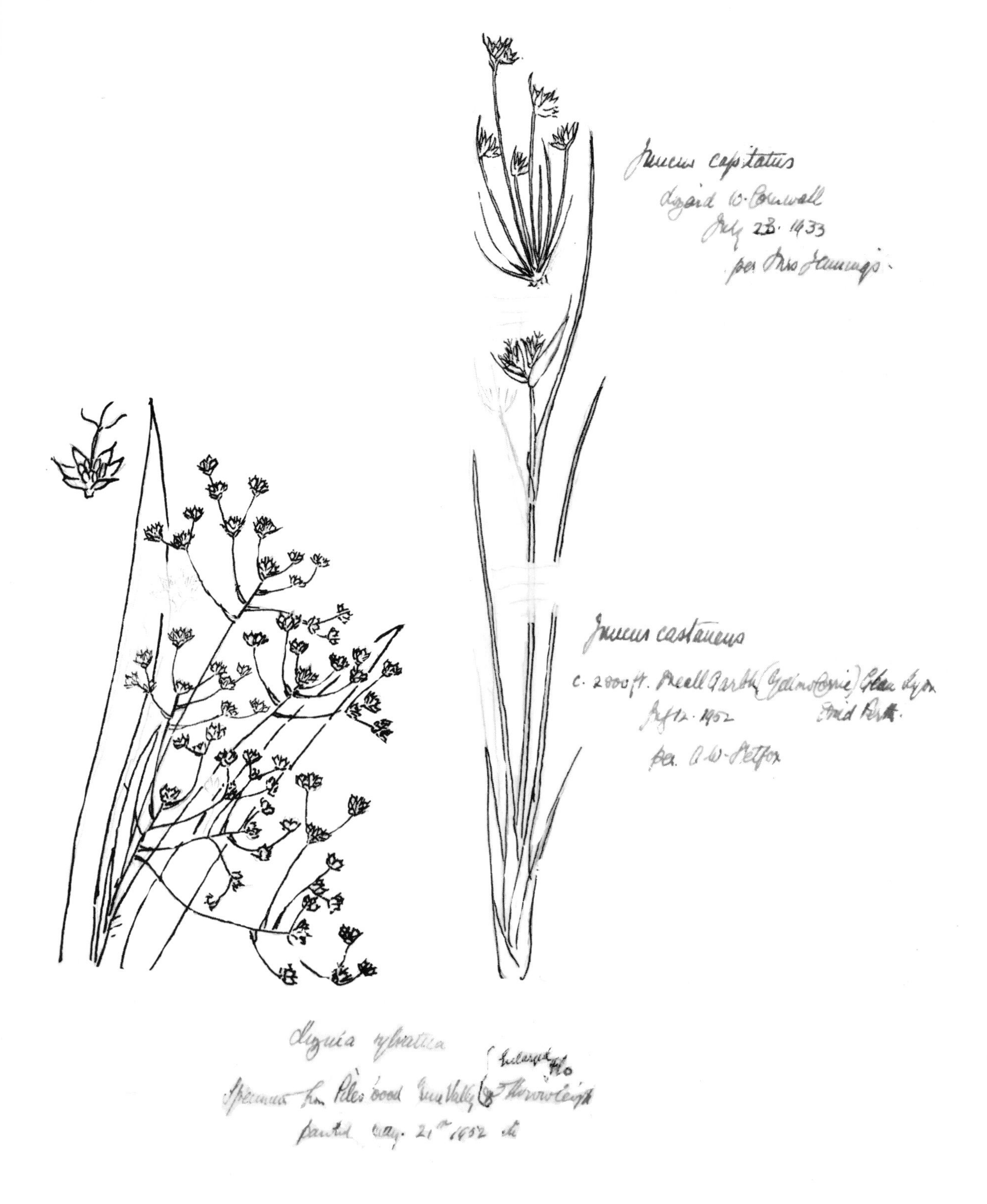

Luzula sylvatica (Huds.) Gaudin.
Great Wood-rush

Juncus capitatus Weigel.
Capitate Rush

Juncus castaneus Sm.
Chestnut Rush

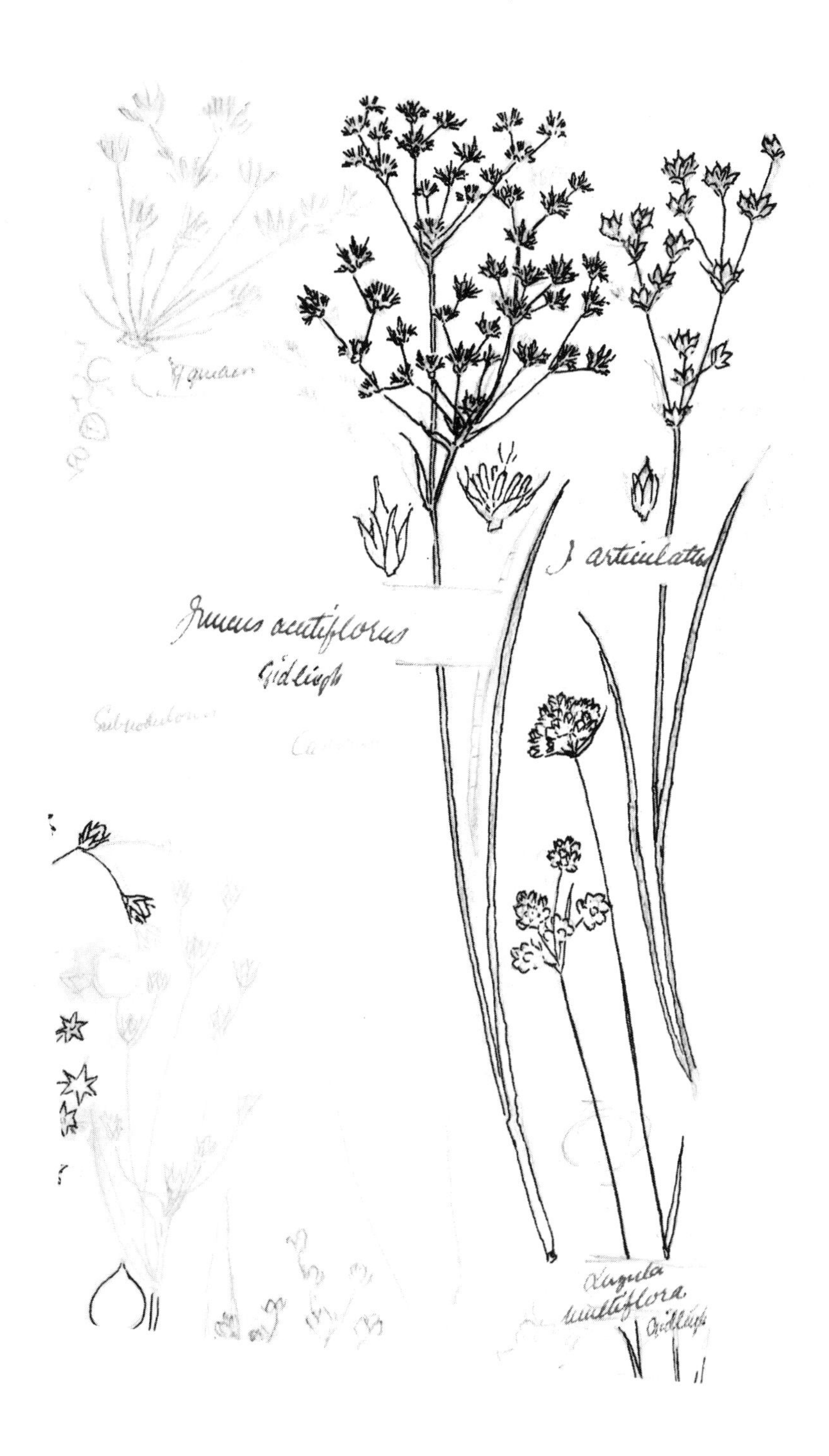

Redrawn from original sketches for the *Flora*

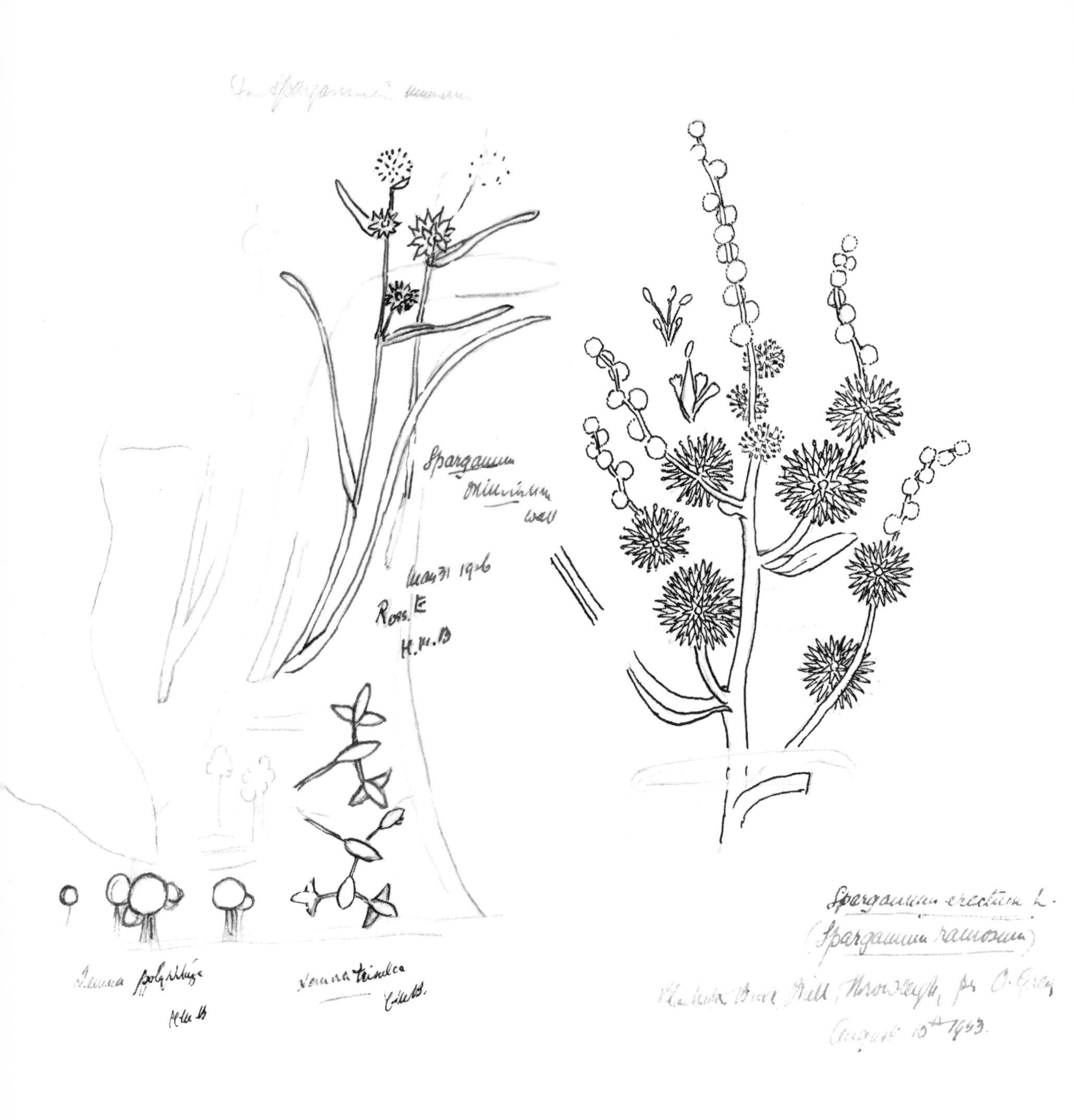

Sparganium minimum Wallr.
Small Bur-reed

Sparganium erectum L.
Branched Bur-reed

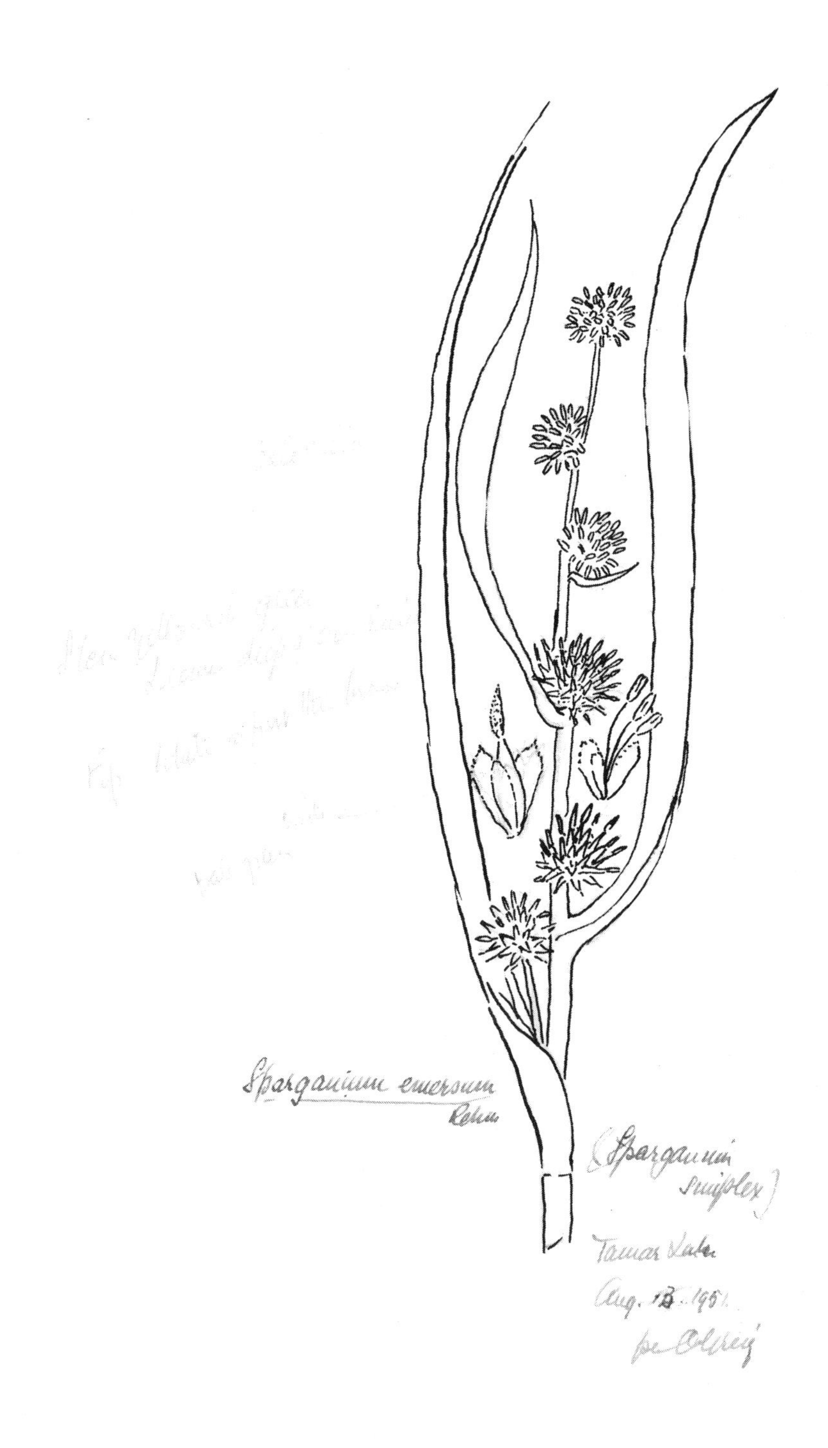

Sparganium emersum Rehm.
Unbranched Bur-reed

Plate 89 POTAMOGETONACEAE

Potamogeton nodosus Poir.
Loddon Pondweed

Potamogeton gramineus L.
Various-leaved Pondweed

Potamogeton perfoliatus L.
Perfoliate Pondweed

Greonlandia densa (L.) Fourr.
Opposite-leaved Pondweed

Redrawn from original sketches for the *Flora*

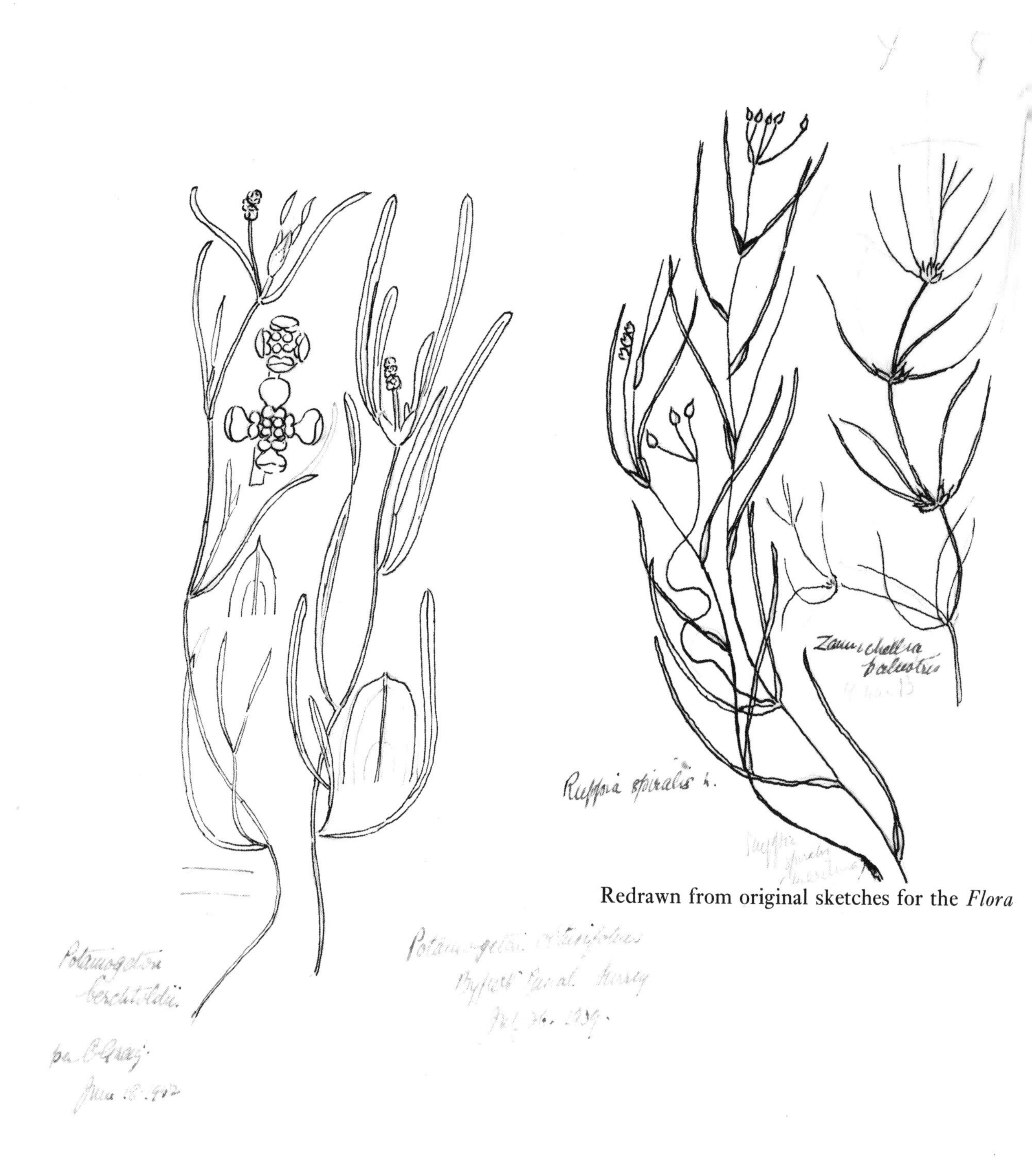

Redrawn from original sketches for the *Flora*

Potamogeton berchtoldii Fieb.
Small Pondweed

Potamogeton obtusifolius Mert. & Koch.
Grassy Pondweed

Redrawn from original sketches for the *Flora*

Plate 91 CYPERACEAE

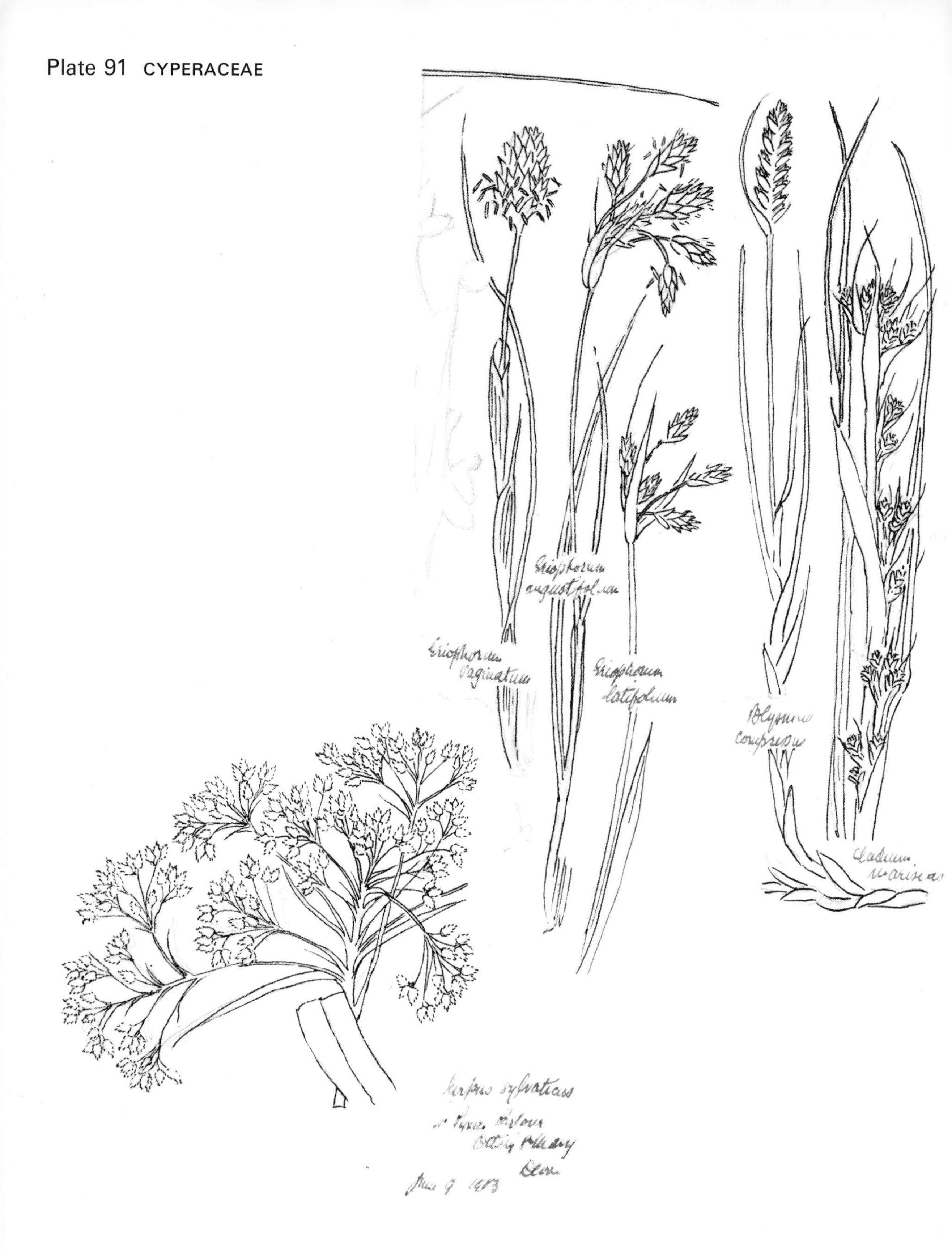

Scirpus sylvaticus L.
Wood Club-rush

Redrawn from original sketches for the *Flora*

Cyperus longus L.
Sweet Galingale

Cyperus fuscus L.
Brown Cyperus

Schoenus nigricans L.
Black Bog-rush

Scirpus lacustris L.
Common Bulrush or Club-rush

Index of Botanical Names

Aceras anthropophorum pl. 81, pl. 82
Aegopodium podagraria pl. 38
Agrimonia eupatoria p. 26
Ajuga chamaepitys pl. 70
Alchemilla alpina pl. 26
Alchemilla filicaulis pl. 26
Allium scorodoprasum pl. 84
Allium triquetrum pl. 85
Alnus glutinosa pl. 77
Althaea hirsuta pl. 18
Althaea officinalis pl. 18
Anemone nemorosa pl. 1
Angelica sylvestris pl. 40
Antennaria dioica pl. 44
Anthyllis vulneraria pl. 23
Aphanes arvensis pl. 26
Apium nodiflorum pl. 37
Aquilegia vulgaris pl. 4
Arabis glabra pl. 7
Arbutus unedo pl. 55
Arctium pubens pl. 47
Arctostaphylos uva-ursi pl. 55
Arenaria ciliata pl. 15
Arenaria norvegica pl. 15
Artemisia absinthium pl. 46
Artemisia campestris pl. 46
Artemisia maritima pl. 46
Artemisia vulgaris pl. 46
Asparagus officinalis pl. 84
Aster tripolium pl. 44
Astragalus alpinus pl. 24
Astragalus glycyphyllos pl. 24
Atriplex hastata pl. 72
Atriplex littoralis pl. 72
Atriplex patula pl. 72
Atropa bella-donna pl. 61

Ballota nigra pl. 70
Barbarea verna pl. 7
Bartsia viscosa pl. 65
Beta maritima pl. 72
Betula pendula pl. 76
Betula pubescens pl. 76
Bidens tripartita pl. 45
Blackstonia perfoliata pl. 58
Brassica oleracea pl. 9
Bryonia dioica pl. 36
Bunium bulbocastanum pl. 37
Butomus umbellatus pl. 79

Cakile maritima pl. 11
Calamintha ascendens pl. 68
Calamintha nepeta pl. 68
Caltha palustris pl. 4
Campanula glomerata pl. 54
Campanula patula pl. 54
Capsella bursa-pastoris pl. 10
Cardamine amara pl. 7
Cardamine bulbifera pl. 8
Cardamine impatiens pl. 8
Cardamine pratensis pl. 7
Cardaminopsis petraea pl. 7
Carduus nutans pl. 48
Carduus tenuiflorus pl. 48
Centaurea aspera pl. 49
Centaurea calcitrapa pl. 49
Centaurea scabiosa pl. 49
Centaurium erythraea pl. 58
Centaurium pulchellum pl. 58
Centaurium scilloides pl. 58
Cephalanthera damasonium p. 80
Cephalanthera longifolia pl. 80
Chaerophyllum temulentum pl. 38
Chenopodium album pl. 71
Chenopodium bonus-henricus pl. 72
Chenopodium ficifolium pl. 72
Chenopodium polyspermum p. 71
Chenopodium rubrum pl. 72
Cicerbita alpina pl. 53
Cicuta virosa pl. 37
Circaea alpina pl. 35
Cirsium acaule pl. 48
Cirsium palustre pl. 48
Cirsium vulgare pl. 48
Clematis vitalba pl. 1
Coeloglossum viride pl. 82
Conium maculatum pl. 36
Convallaria majalis pl. 84
Corrigiola litoralis pl. 71
Corydalis lutea pl. 6
Crambe maritima pl. 11
Crepis biennis pl. 50
Crepis foetida pl. 50
Crepis mollis pl. 50
Crinitaria linosyris pl. 44
Cuscuta europaea pl. 61
Cynoglossum germanicum pl. 59
Cyperus fuscus pl. 91
Cyperus longus pl. 91
Cypripedium calceolus pl. 82

Dactylorhiza incarnata pl. 81
Descurainia sophia pl. 8
Dianthus deltoides pl. 13
Dipsacus fullonum pl. 43
Dipsacus pilosus pl. 43
Drosera intermedia pl. 34
Drosera rotundifolia pl. 34
Dryas octopetala pl. 29

Echium vulgare pl. 61
Elodea canadensis pl. 79
Epilobium anagallidifolium pl. 35
Epilobium angustifolium pl. 35
Epilobium hirsutum pl. 35
Epilobium parviflorum pl. 35
Epilobium roseum pl. 35
Epipactis helleborine pl. 80
Epipactis leptochila pl. 80
Epipactis palustris pl. 81
Erica ciliaris pl. 55
Erica vagans pl. 55
Erodium cicutarium pl. 20
Erodium moschatum pl. 20
Eryngium maritimum pl. 36
Erysimum cheiranthoides pl. 9
Euonymus europaeus pl. 20

Eupatorium cannabinum pl. 44
Euphorbia helioscopia pl. 75
Euphorbia hyberna pl. 75
Euphorbia paralias pl. 75
Euphorbia peplis pl. 75
Euphorbia platyphyllos pl. 75
Euphrasia brevipila pl. 64

Foeniculum vulgare pl. 39
Fragaria vesca pl. 27

Galium album pl. 42
Galium boreale pl. 42
Galium debile pl. 42
Galium odoratum pl. 42
Galium saxatile pl. 42
Gentiana anglica pl. 59
Gentiana pneumonanthe pl. 59
Gentianella campestris pl. 59
Geranium pusillum pl. 19
Geranium sanguineum pl. 19
Geranium sylvaticum pl. 19
Geum rivale pl. 29
Glaucium flavum pl. 5
Goodyera repens pl. 80
Groenlandia densa pl. 89
Gymnadenia conopsea pl. 82

Halimione portulacoides pl. 72
Helianthemum appenninum pl. 11
Helianthemum canum pl. 11
Helianthemum nummularium pl. 11
Helleborus viridis pl. 4
Heracleum sphondylium pl. 40
Hesperis matronalis pl. 8
Himantoglossum hircinum pl. 81
Hippuris vulgaris pl. 34
Honkenya peploides pl. 16
Hypericum calycinum pl. 17
Hypericum linarifolium pl. 17
Hypericum perforatum pl. 17
Hypericum undulatum pl. 17

Iberis amara pl. 10
Impatiens noli-tangere pl. 20
Inula helenium pl. 45
Iris foetidissima pl. 83

Juncus acutus pl. 87
Juncus capitatus pl. 87
Juncus castaneus pl. 87
Juncus compressus pl. 86
Juncus conglomeratus pl. 86
Juncus effusus pl. 86
Juncus gerardii pl. 86
Juncus maritimus pl. 87

Lamium album pl. 70
Lamium amplexicaule pl. 70
Lamium hybridum pl. 70
Lamium purpureum pl. 70
Lapsana communis pl. 50
Lathyrus aphaca pl. 25
Lathyrus sylvestris pl. 25
Lavatera arborea pl. 18
Legousia hybrida pl. 54
Leontodon autumnalis pl. 53
Leontodon hispidus pl. 53
Leontodon taraxacoides pl. 53
Lepidium latifolium pl. 10
Leucorchis albida pl. 82
Ligusticum scoticum pl. 40
Lilium pyrenaicum pl. 85
Limonium bellidifolium pl. 56
Limonium transwallianum pl. 56
Linaria purpurea pl. 62
Linaria vulgaris pl. 62
Linnaea borealis pl. 41
Linum perenne pl. 19
Liparis loeselii pl. 80
Lobelia urens pl. 54
Lotus corniculatus pl. 23
Lotus siliquosus pl. 23
Lotus tenuis pl. 23
Luzula forsteri pl. 87
Luzula pilosa pl. 87
Luzula sylvatica pl. 87
Lychnis flos-cuculi pl. 14
Lycopsis arvensis pl. 60
Lysimachia nemorum pl. 57
Lysimachia nummularia pl. 57
Lysimachia vulgaris pl. 57
Lythrum salicaria pl. 34

Maianthemum bifolium pl. 84
Malva moschata pl. 18
Matricaria matricarioides pl. 46
Matthiola sinuta pl. 6
Meconopsis cambrica pl. 5
Medicago sativa pl. 21
Melampyrum arvense pl. 65
Melilotus altissima pl. 22
Melittis melissophyllum pl. 70
Mentha × gentilis pl. 67
Mentha × piperita pl. 67
Mentha rotundifolia pl. 67
Mertensia maritima pl. 60
Moehringia trinervia pl. 15
Moneses uniflora pl. 56
Myosotis alpestris pl. 60
Myosotis caespitosa pl. 60
Myosotis ramosissima pl. 60
Myosotis secunda pl. 60
Myosotis stolonifera pl. 60
Myriophyllum verticillatum pl. 34
Myrrhis odorata pl. 38

Nasturtium officinale pl. 7
Nasturtium palustre pl. 7
Nepeta cataria pl. 68

Oenanthe crocata pl. 39
Oenanthe pimpinelloides pl. 39
Onobrychis viciifolia pl. 24
Ononis repens pl. 21
Onopordon acanthium pl. 49
Ornithogalum nutans pl. 85
Orobanche elatior pl. 66
Orobanche rapum-genistae pl. 66
Otanthus maritimus pl. 45
Oxytropis campestris pl. 23
Oxytropis halleri pl. 23

Paeonia mascula pl. 4
Papaver dubium pl. 5
Papaver somniferum pl. 5
Parietaria diffusa pl. 76
Pedicularis palustris pl. 65
Pentaglottis sempervirens pl. 60
Physospermum cornubiense pl. 36
Picris hieracioides pl. 50
Pimpinella saxifraga pl. 38
Platanthera bifolia pl. 82
Polycarpon tetraphyllum pl. 16
Polygonum amphibium pl. 73
Polygonum bistorta pl. 73
Polygonum hydropiper pl. 73
Polygonum lapathifolium pl. 73
Polygonum nodosum pl. 73
Polygonum persicaria pl. 73
Polygonum viviparum pl. 73
Populus alba pl. 78
Populus canescens pl. 78
Potamogeton berchtoldii pl. 90
Potamogeton gramineus pl. 89
Potamogeton nodosus pl. 89
Potamogeton obtusifolius pl. 90
Potamogeton perfoliatus pl. 89
Potentilla anserina pl. 27
Potentilla erecta pl. 27
Potentilla norvegica pl. 27
Potentilla palustris pl. 27
Potentilla reptans pl. 27
Potentilla rupestris pl. 27

Potentilla sterilis pl. 27
Primula elatior pl. 57
Primula farinosa pl. 57
Primula scotica pl. 57
Primula veris pl. 57
Primula vulgaris pl. 57
Prunella vulgaris pl. 69
Prunus avium pl. 26
Prunus cerasus pl. 26
Prunus domestica pl. 26
Prunus insititia pl. 26
Prunus spinosa pl. 26
Pulicaria dysenterica pl. 45
Pulicaria vulgaris pl. 45
Pulsatilla vulgaris pl. 1
Pyrola rotundifolia pl. 56
Pyrus pyraster pl. 31

Ranunculus baudotii pl. 2
Ranunculus bulbosus pl. 3
Rananculus ficaria pl. 3
Ranunculus fluitans pl. 2
Ranunculus lingua pl. 3
Ranunculus peltatus pl. 2
Ranunculus repens pl. 3
Ranunculus tripartitus pl. 2
Raphanus maritimus pl. 11
Reseda lutea pl. 11
Reseda luteola pl. 11
Rhinanthus serotinus pl. 65
Rhodiola rosea pl. 33
Rhynchosinapis monensis pl. 9
Ribes spicatum pl. 33
Romulea columnae pl. 83
Rorippa amphibia pl. 7
Rorippa palustris pl. 7
Rorippa sylvestris pl. 7
Rosa canina pl. 30
Rosa tomentosa pl. 30
Rosa villosa pl. 30
Rubia peregrina pl. 42
Rubus bellardii pl. 29
Rubus caesius pl. 29
Rubus idaeus pl. 28
Rubus radula pl. 28
Rubus sylvaticus pl. 28
Rubus vestitus pl. 28
Rubus villicanlis pl. 28
Rumex acetosa pl. 74
Rumex acetosella pl. 74
Rumex conglomeratus pl. 74
Rumex longifolius pl. 74
Rumex maritimus pl. 74
Rumex obtusifolius pl. 74
Rumex rupestris pl. 74
Ruscus aculeatus pl. 84

Sagina procumbens pl. 16
Sagittaria sagittifolia pl. 79
Salix arbuscula pl. 78
Salix caprea pl. 78
Salix fragilis pl. 77
Salix lapponum pl. 77
Salix myrsinites pl. 78
Salix nigricans pl. 78
Salix triandra pl. 77
Salvia pratensis pl. 68
Sambucus ebulus pl. 41
Saponaria officinalis pl. 13
Saxifraga cespitosa pl. 32
Saxifraga granulata pl. 32
Saxifraga hypnoides pl. 32
Saxifraga oppositifolia pl. 32
Saxifraga × urbium pl. 32
Schoenus nigricans pl. 91
Scirpus lacustris pl. 91
Scirpus sylvaticus pl. 91
Scrophularia aquatica pl. 63
Scrophularia scorodonia pl. 63
Scutellaria galericulata pl. 68
Sedum forsteranum pl. 33
Senecio erucifolius pl. 47
Senecio spathulifolius pl. 47
Senecio sylvaticus pl. 47
Serratula tinctoria pl. 49
Silene alba pl. 14
Silene conica pl. 13
Silene dioica pl. 14
Silene noctiflora pl. 14
Silene nutans pl. 13
Silene otites pl. 13
Simethis planifolia pl. 84
Sinapis arvensis pl. 9
Sium latifolium pl. 37
Solanum dulcamara pl. 61
Solanum nigrum pl. 61
Solidago virgaurea pl. 44
Sonchus arvensis pl. 53
Sonchus palustris pl. 53
Sorbus aria pl. 31
Sorbus aucuparia pl. 31
Sorbus latifolia pl. 31
Sorbus torminalis pl. 31
Sparganium emersum pl. 88
Sparganium erectum pl. 88
Sparganium minimum pl. 88
Spergularia media pl. 16
Spergularia rupicola pl. 16
Stachys alpina pl. 69
Stachys palustris pl. 69
Stellaria holostea pl. 15
Stratiotes aloides pl. 79
Swida sanguinea pl. 41
Symphytum officinale pl. 60
Symphytum tuberosum pl. 60

Teucrium scordium pl. 70
Teucrium scorodonia pl. 70
Thalictrum minus pl. 1
Thlaspi perfoliatum pl. 10
Tilia cordata pl. 18
Torilis arvensis pl. 40
Trifolium campestre pl. 23
Trifolium fragiferum pl. 23
Trifolium micranthum pl. 23
Trifolium molinerii pl. 22
Tripleurospermum maritimum pl. 46
Tulipa sylvestris pl. 85

Ulex europaeus pl. 21
Ulex minor pl. 21
Ulmus glabra pl. 76
Ulmus procera pl. 76
Umbilicus rupestris pl. 33
Urtica dioica pl. 76
Utricularia intermedia pl. 66

Vaccaria pyramidata pl. 13
Vaccinium myrtillus pl. 55
Valeriana dioica pl. 43
Valeriana officinalis pl. 43
Verbascum nigrum pl. 62
Verbascum thapsus pl. 62
Veronica beccabunga pl. 64
Veronica fruticans pl. 64
Veronica persica pl. 63
Veronica polita pl. 63
Veronica spicata pl. 64
Veronica verna pl. 63
Viburnum opulus pl. 41
Vicia angustifolia pl. 25
Vicia hirsuta pl. 24
Vicia lutea pl. 25
Vicia orobus pl. 24
Viola lactea pl. 12
Viola odorata pl. 12
Viola reichenbachiana pl. 12
Viola tricolor pl. 12

Index of Common English Names

Agrimony, Common see *Agrimonia eupatoria* pl. 26
Agrimony, Hemp see *Eupatorium cannabinum* pl. 44
Alder see *Alnus glutinosa* pl. 77
Alkanet see *Pentaglottis sempervirens* pl. 60
Allseed, Four-leaved see *Polycarpon tetraphyllum* pl. 16
Almond Willow see *Salix triandra* pl. 77
Alpine Bistort see *Polygonum viviparum* pl. 73
Alpine Enchanter's Nightshade see *Circaea alpina* pl. 35
Alpine Forget-me-not see *Myosotis alpestris* pl. 60
Alpine Lady's Mantle see *Alchemilla alpina* pl. 26
Alpine Lettuce see *Cicerbita alpina* pl. 53
Alpine Milk-vetch see *Astragalus alpinus* pl. 24
Alpine Willow Herb see *Epilobium anagallidifolium* pl. 35
Alpine Woundwort see *Stachys alpina* pl. 69
American Cress see *Barbarea verna* pl. 7
Amphibious Periscaria see *Polygonum amphibium* pl. 73
Amphibious Yellow Cress see *Rorippa amphibia* pl. 7
Anemone, Wood see *Anemone nemorosa* pl. 1
Angelica, Wild see *Angelica sylvestris* pl. 40
Arctic Sandwort see *Arenaria norvegica* pl. 15
Arrowhead see *Sagittaria sagittifolia* pl. 79
Ash, Mountain see *Sorbus aucuparia* pl. 31
Asparagus, Wild see *Asparagus officinalis* pl. 84
Aster, Sea, see *Aster tripolium* pl. 44
Avens, Mountain see *Dryas octopetala* pl. 29
Avens, Water see *Geum rivale* pl. 29

Balm, Bastard see *Melittis melissophyllum* pl. 69
Balm-leaved Figwort see *Scrophularia scorodonia* pl. 63
Balsam, Wild see *Impatiens noli-tangere* pl. 20
Barren Strawberry see *Potentilla sterilis* pl. 27
Bartsia, Yellow see *Parentucellia viscosa* pl. 65
Bastard Balm see *Melittis melissophyllum* pl. 70
Bay, Rose see *Epilobium angustifolium* pl. 35
Beam, White see *Sorbus aria* pl. 31
Bear Berry see *Arctostaphylos uva-ursi* pl. 55
Bedstraw, Heath see *Galium saxatile* pl. 42
Bedstraw, Hedge see *Galium album* pl. 42
Bedstraw, Northern see *Galium boreale* pl. 42
Beet, Sea see *Beta vulgaris* pl. 72
Bell-flower, Clustered see *Campanula glomerata* pl. 54
Bell-flower, Spreading see *Campanula patula* pl. 54
Berry, Bear see *Arctostaphylos uva-ursi* pl. 55
Bilberry see *Vaccinium myrtillus* pl. 55
Birch, Brown see *Betula pubescens* pl. 76
Birch, Silver see *Betula pendula* pl. 76
Bird Cherry see *Prunus padus* pl. 26
Bird's-eye Primrose see *Primula farinosa* pl. 57
Bird's-foot Trefoil see *Lotus corniculatus* pl. 23
Bird's-foot Trefoil, Marsh see *Lotus uliginosus* pl. 23
Bird's-foot Trefoil, Narrow-leaved see *Lotus tenuis* pl. 23
Bistort see *Polygonum bistorta* pl. 73
Bistort, Alpine see *Polygonum viviparum* pl. 73
Biting Persicaria see *Polygonum hydropiper* pl. 73
Bitter Cress, Large, see *Cardamine amara* pl. 7
Bitter Cress, Narrow-leaved see *Cardamine impatiens* pl. 8
Black Bog-rush see *Schoenus nigricans* pl. 91
Black Nightshade see *Solanum nigrum* pl. 61
Blackthorn see *Prunus spinosa* pl. 26
Bladder Sedge see *Carex vesicaria* pl. 94
Blood-red Geranium see *Geranium sanguineum* pl. 19
Blue Flax see *Linum perenne* pl. 19
Blue Lobelia see *Lobelia urens* pl. 54
Bog-rush, Black see *Schoenus nigricans* pl. 91
Branched Bur-reed see *Sparganium erectum* pl. 88
Breckland Wormwood see *Artemisia campestris* pl. 46
Broad Helleborine see *Epipactis helleborine* pl. 80

Broad-leaved Dock see *Rumex obtusifolius* pl. 74
Broad-leaved Spurge see *Euphorbia platyphyllos* pl. 75
Broad-leaved White Bean see *Sorbus latifolia* pl. 31
Brooklime see *Veronica beccabunga* pl. 64
Broom, Butcher's see *Ruscus aculeatus* pl. 84
Broomrape, Greater see *Orobanche rapumgenistae* pl. 66
Broomrape, Tall see *Orobanche elatior* pl. 66
Brown Birch see *Betula pubescens* pl. 76
Brown Cyperus see *Cyperus fuscus* pl. 91
Bryony, White see *Bryonia dioica* pl. 36
Bugloss, Lesser see *Lycopsis arvensis* pl. 60
Bugloss, Viper's see *Echium vulgare* pl. 61
Bulbous Buttercup see *Ranunculus bulbosus* pl. 3
Bulbous Saxifrage see *Saxifraga granulata* pl. 32
Bullace see *Prunus domestica* subsp. *institia* pl. 26
Bulrush, Common see *Scirpus lacustris* pl. 91
Bur-Marigold see *Bidens tripartita* pl. 45
Bur-reed, Branched see *Sparganium erectum* pl. 88
Bur-reed, Small see *Sparganium minimum* pl. 88
Bur-reed, Unbranched see *Sparganium emersum* pl. 88
Burdock, Common see *Arctium pubens* pl. 47
Burnet Saxifrage see *Pimpinella saxifraga* pl. 38
Butcher's Broom see *Ruscus aculeatus* pl. 84
Buttercup, Bulbous see *Ranunculus bulbosus* pl. 3
Buttercup, Creeping see *Ranunculus repens* pl. 3
Butterfly Orchid, Lesser see *Platanthera bifolia* pl. 82

Cabbage, Isle of Man see *Rhynchosinapis monensis* pl. 9
Cabbage, Wild see *Brassica oleracea* pl. 9
Calamint, Common see *Calamintha ascendens* pl. 68
Calamint, Lesser see *Calamintha nepeta* pl. 68
Campion, Red see *Silene dioica* pl. 14
Campion, White see *Silene alba* pl. 14
Canadian Pondweed see *Elodea canadensis* pl. 79
Candytuft see *Iberis amara* pl. 10
Capitate Rush see *Juncus capitatus* pl. 87
Caraway, Tuberous see *Bunium bulbocastanum* pl. 37
Cat Mint see *Nepeta cataria* pl. 68
Catchfly, Night-flowering see *Silene noctiflora* pl. 14
Catchfly, Nodding see *Silene nutans* pl. 13
Catchfly, Nottingham see *Silene nutans* pl. 13
Catchfly, Spanish see *Silene otites* pl. 13
Catchfly, Striated see *Silene conica* pl. 13
Cat's-foot see *Antennaria dioica* pl. 44
Cat's-tail, Lesser see *Phleum bertolonii* pl. 95
Celandine, Lesser see *Ranunculus ficaria* pl. 3
Centaury, Common see *Centaurium erythraea* pl. 58
Centaury, Lesser see *Centaurium pulchellum* pl. 58
Centaury, Perennial see *centaurium scilloides* pl. 58
Chamomile, Scentless see *Tripleurospermum maritimum* pl. 46
Charlock see *Sinapis arvensis* pl. 9
Cherry, Bird see *Prunus padus* pl. 26
Cherry, Sour see *Prunus cerasus* pl. 26
Cherry, Wild see *Prunus avium* pl. 26
Chervil see *Chaerophyllum temulentum* pl. 38
Chestnut Rush see *Juncus casteneus* pl. 87
Cicely, Sweet see *Myrrhis odorata* pl. 38
Ciliate Heath see *Erica ciliaris* pl. 55
Cinquefoil, Creeping see *Potentilla reptans* pl. 27
Cinquefoil, Marsh see *Potentilla palustris* pl. 27
Cinquefoil, Rock see *Potentilla rupestris* pl. 27
Clary, Meadow see *Salvia pratensis* pl. 68
Cliff Meadow Rue see *Thalictrum minus* pl. 1
Clover, Large Lizard see *Trifolium molinerii* pl. 22
Clover, Strawberry-headed see *Trifolium fragiferum* pl. 23
Club-rush, Round-headed see *Scirpus holoschoenus* pl. 91
Club-rush, Wood see *Scirpus sylvaticus* pl. 91
Clustered Bell-flower see *Campanula glomerata* pl. 54
Columbine see *Aquilegia vulgaris* pl. 4
Comfrey see *Symphytum officinale* pl. 60
Comfrey, Tuberous see *Symphytum tuberosum* pl. 60
Common Agrimony see *Agrimonia eupatoria* pl. 26
Common Bulrush see *Scirpus lacustris* pl. 91
Common Burdock see *Aretium pubens* pl. 47
Common Calamint see *Calamintha ascendens* pl. 68
Common Centaury see *Centaurium erythraea* pl. 58
Common Gorse see *Ulex europaeus* pl. 21
Common Mullein see *Verbascum thapsus* pl. 62
Common Orache see *Atriplex patula* pl. 72
Common Pearlwort see *Sagina procumbens* pl. 16
Common Persicaria see *Polygonum persicaria* pl. 73
Common Rockrose see *Helianthemum nummularium* pl. 11
Common Rush see *Juncus subuliflorus* pl. 86
Common Sorrel see *Rumese acetosa* pl. 74

Common Storksbill see *Erodium cicutarium* pl. 20
Common Teasel see *Dipsacus fullonum* pl. 43
Common Toadflax see *Linaria vulgaris* pl. 62
Common Tormentil see *Potentilla erecta* pl. 27
Common Yellow Rocket see *Barbarea vulgaris* pl. 7
Coral Root see *Cardamine bulbifera* pl. 8
Corky-fruited Water Dropwort see *Oenanthe pimpinelloides* pl. 39
Corn Sowthistle see *Sonchus arvensis* pl. 53
Cornel see *Swida sanguinea* pl. 41
Cornish Heath see *Erica vagans* pl. 55
Corydalis, Yellow see *Corydalis lutea* pl. 6
Cow Parsnip see *Heracleum sphondylium* pl. 40
Cow-wheat, Field see *Melampyrum arvense* pl. 65
Cowbane see *Cicuta virosa* pl. 37
Cowslip see *Primula veris* pl. 57
Crack Willow see *Salix fragilis* pl. 77
Cranesbill, Small-flowered see *Geranium pusillum* pl. 19
Cranesbill, Wood see *Geranium sylvaticum* pl. 19
Creeping Buttercup see *Ranunculus repens* pl. 3
Creeping Cinquefoil see *Potentilla reptans* pl. 27
Creeping Jenny see *Lysimachia numularia* pl. 57
Creeping Lady's Tresses see *Goodyera repens* pl. 80
Creeping Yellow Cress see *Rorippa sylvestris* pl. 7
Cress, American see *Barbarea verna* pl. 7
Cress, Amphibious Yellow see *Rorippa amphibia* pl. 7
Cress, Creeping Yellow see *Rorippa sylvestris* pl. 7
Cress, Glabrous Tower- see *Arabis glabra* pl. 7
Cress, Large Bitter see *Cardamine amara* pl. 7
Cress, Marsh Yellow see *Rorippa palustris* pl. 7
Cress, Mountain Rock- see *Cardaminopsis petraea* pl. 7
Cress, Narrow-leaved Bitter see *Cardamine impatiens* pl. 8
Cress, Perfoliate Penny see *Thlaspi perfoliatum* pl. 10
Cress, Water see *Nasturtium officinale* p. 7
Cress, Winter see *Barbarea vulgaris* pl. 7
Crocus, Warren see *Romulea columnae* pl. 83
Crowfoot, Little Three-lobed see *Ranunculus tripartitus* pl. 2
Crowfoot, River see *Ranunculus fluitans* pl. 2
Crowfoot, Seaside see *Ranunculus baudotti* pl. 2
Cuckoo Flower see *Cardamine pratensis* pl. 7
Currant, Upright Red see *Ribes spicatum* pl. 33
Cut-leaved Dead-nettle see *Lamium hybridum* pl. 70
Cyperus, Brown see *Cyperus fuscus* pl. 91

Dame's Violet see *Hesperis matronalis* pl. 8
Danewort see *Sambucus ebulus* pl. 41
Dark Mullein see *Verbascum nigrum* pl. 62
Dark-leaved Willow see *Salix nigricans* pl. 78
Dead-nettle, Cut-leaved see *Lamium hybridum* pl. 70
Dead-nettle, Purple see *Lamium purpureum* pl. 70
Dead-nettle, White see *Lamium album* pl. 70
Deadly Nightshade see *Atropa belladonna* pl. 61
Dewberry see *Rubus caesius* pl. 29
Dittander see *Lepidium latifolium* pl. 10
Dock, Broad-leaved see *Rumex obtusifolius* pl. 74
Dock, Golden see *Rumex maritimus* pl. 74
Dock, Long-leaved see *Rumex longifolius* pl. 74
Dock, Sharp see *Rumex conglomeratus* pl. 74
Dock, Shore see *Rumex rupestris* pl. 74
Dodder, Greater see *Cuscuta europaea* pl. 61
Dog Rose see *Rosa canina* pl. 30
Dogwood see *Swida sanguinea* pl. 41
Downy Rose see *Rosa tomentosa* pl. 30
Downy Willow see *Salix lapponum* pl. 77
Drooping Star of Bethlehem see *Ornithogalum nutans* pl. 85
Dropwort, Corky-fruited Water see *Oenanthe pimpinelloides* pl. 39
Dropwort, Hemlock Water see *Oenanthe crocata* pl. 39
Dwarf English Felwort see *Gentianella anglica* pl. 59
Dyer's Rocket see *Reseda luteola* pl. 11

Early Forget-me-not see *Myosotis ramosissima* pl. 60
Elder, Ground see *Aegopodium podagraria* pl. 38
Elecampane see *Inula helenium* pl. 45
Elm, English see *Ulmus procera* pl. 76
Elm, Wych see *Ulmus glabra* pl. 76
Enchanter's Nightshade, Alpine see *Circaea alpina* pl. 35
English Elm see *Ulmus procera* pl. 76
English Felwort, Dwarf see *Gentianella anglica* pl. 59
Eyebright, Short-haired see *Euphrasia brevipila* pl. 64

Fat Hen see *Chenopodium album* pl. 71
Felwort, Dwarf English see *Gentianella anglica* pl. 59
Felwort, Field see *Gentianella campestris* pl. 59
Fen Orchid see *Liparis loeselii* pl. 80
Fen Sowthistle see *Sonchus palustris* pl. 53
Fennell see *Foeniculum vulgare* pl. 39
Field Cow-wheat see *Melampyrum arvense* pl. 65

Field Felwort see *Gentianella campestris* pl. 59
Fig-leaved Goose-foot see *Chenopodium ficifolium* pl. 72
Figwort, Balm-leaved see *Scrophularia scorodonia* pl. 63
Figwort, Water see *Scrophularia auriculata* pl. 63
Flax, Blue see *Linum perenne* pl. 19
Fleabane see *Pulicaria dysenterica* pl. 45
Fleabane, Lesser see *Pulicaria vulgaris* pl. 45
Fleawort, Spathulate see *Senecio spathulifolius* pl. 47
Flixweed see *Descurainia sophia* pl. 8
Flowering Rush see *Butomus umbellatus* pl. 79
Forget-me-not, Alpine see *Myosotis alpestris* pl. 60
Forget-me-not, Early see *Myosotis ramosissima* pl. 60
Forget-me-not, Lesser Water see *Myosotis caespitosa* pl. 60
Forget-me-not, Marsh see *Myosotis secunda* pl. 60
Forget-me-not, Short-leaved see *Myosotis stolonifera* pl. 60
Forget-me-not, Water see *Myosotis scorpioides* pl. 60
Forster's Wood-rush see *Luzula forsteri* pl. 87
Four-leaved Allseed see *Polycarpon tetraphyllum* pl. 16
Frog Orchid see *Coeloglossum viride* pl. 82
Furze, see *Ulex europaeus* pl. 21
Furze, Small see *Ulex minor* pl. 21

Galingale, Sweet see *Cyperus longus* pl. 91
Garlic, Triangular-stalked see *Allium triquetrum* pl. 85
Gean see *Prunus avium* pl. 26
Gentian, Marsh see *Gentiana pneumonanthe* pl. 59
Geranium, Blood-red see *Geranium sanguineum* pl. 19
Germander, Water see *Teucrium scordium* pl. 70
Glabrous Tower-Cress see *Arabis glabra* pl. 7
Golden Dock see *Rumex maritimus* pl 74
Golden Rod see *Solidago virgaurea* pl. 44
Goldilocks see *Crinitaria linosyris* pl. 44
Good King Henry see *Chenopodium bonushenricus* pl. 72
Goose-foot, Fig-leaved see *Chenopodium ficifolium* pl. 72
Goose-foot, Many-seeded see *Chenopodium polyspermum* pl. 71
Goose-foot, Red see *Chenopodium rubrum* pl. 72
Gorse, Common see *Ulex europaeus* pl. 21
Goutweed see *Aegopodium podograria* pl. 38
Grass-leaved Orache see *Atriplex littoralis* pl. 72
Grassy Pondweed see *Potamogeton obtusifolius* pl. 90
Great Knapweed see *Centaurea scabiosa* pl. 49
Great Sallow see *Salix caprea* pl. 78
Great Spearwort see *Ranunculus reptans* pl. 3
Great Willow Herb see *Epilobium hirsutum* pl. 35
Great Wood-rush see *Luzula sylvatica* pl. 87
Greater Broomrape see *Orobanche rapumgenistae* pl. 66
Greater Dodder see *Cuscuta europaea* pl. 61
Greater Hawk's-beard see *Crepis biennis* pl. 50
Greater Hayrattle see *Rhinanthus serotinus* pl. 65
Greater Stitchwort see *Stellaria holostea* pl. 15
Green Hellebore see *Helleborus viridis* pl. 4
Green Hound's Tongue see *Cynoglossum germanicum* pl. 59
Grey Poplar see *Populus canescens* pl. 78
Grey Speedwell see *Veronica polita* pl. 63
Ground Elder see *Aegopodium podagraria* pl. 38
Ground Thistle see *Cirsium acaule* pl. 48
Ground-pine see *Ajuga chamaepitys* pl. 70
Groundsel, Wood see *Senecio sylvaticus* pl. 47
Guelder Rose see *Viburnum opulus* pl. 41

Hairy Tare see *Vicia hirsuta* pl. 24
Hairy Willow Herb see *Epilobium parviflorum* pl. 35
Hairy Wood-rush see *Luzula pilosa* pl. 87
Hastate Orache see *Atriplex hastata* pl. 72
Hawkbit see *Leontodon taraxacoides* pl. 53
Hawkbit, Rough see *Leontodon hispidus* pl. 53
Hawkbit, Smooth see *Leontodon autumnalis* pl. 53
Hawk's-beard, Greater see *Crepis biennis* pl. 50
Hawk's-beard, Rough see *Crepis biennis* pl. 50
Hawk's-beard, Soft see *Crepis mollis* pl. 50
Hawk's-beard, Stinking see *Crepis foetida* pl. 50
Hawkweed Ox-tongue see *Picris hieracioides* pl. 50
Hayrattle, Greater see *Rhinanthus serotinus* pl. 65
Heath Bedstraw see *Galium saxatile* pl. 42
Heath, Ciliate see *Erica ciliaris* pl. 55
Heath, Cornish see *Erica vagans* pl. 55
Heath Violet, Pale see *Viola lactea* pl. 12
Hedge Bedstraw see *Galium album* pl. 42
Hedge Parsley, Spreading see *Torilis arvensis* pl. 40
Hellebore, Green see *Helleborus viridis* pl. 4
Helleborine, Broad see *Epipactis helleborine* pl. 80
Helleborine, Long-leaved see *Cephalanthera longifolia* pl. 80
Helleborine, Marsh see *Epipactis palustris* pl. 81
Helleborine, Narrow-lipped see *Epipactis leptochila* pl. 80
Helleborine, White see *Cephalanthera damasonium* pl. 80
Hemlock see *Conium maculatum* pl. 36

Hemlock Water Dropwort see *Oenanthe crocata* pl. 39
Hemp Agrimony see *Eupatorium cannabinum* pl. 44
Hen, Fat see *Chenopodium album* pl. 71
Henbit see *Lamium amplexicaule* pl. 70
Henry, Good King see *Chenopodium bonushenricus* pl. 72
Herb, Great Willow see *Epilobium hirsutum* pl. 35
Herb, Hairy Willow see *Epilobium parviflorum* pl. 35
Herb, Pedicelled Willow see *Epilobium roseum* pl. 35
Hispid Marsh Mallow see *Althaea hirsuta* pl. 18
Hoary Ragwort see *Senecio erucifolius* pl. 47
Hoary Rockrose see *Heliathemum canum* pl. 11
Hogweed see *Heracleum sphondylium* pl. 40
Holly, Sea see *Eryngium maritimum* pl. 36
Hop Trefoil see *Trifolium campestre* pl. 23
Horehound see *Ballota nigra* pl. 70
Horned-Poppy, Yellow see *Glaucium flavum* pl. 5
Hound's Tongue see *Cynoglossum officinale* pl. 59

Iris, Stinking see *Iris foetidissima* pl. 83
Irish Sandwort see *Arenaria ciliata* pl. 15
Irish Spurge see *Euphorbia hyberna* pl. 75
Isle of Man Cabbage see *Rhynchosinapis monensis* pl. 9

Jenny, Creeping see *Lysimachia nummularia* pl. 57
Joy, Traveller's see *Clematis vitalba* pl. 1

Kale, Sea see *Crambe maritima* pl. 11
Kidney Vetch see *Anthyllis vulneraria* pl. 23
King Henry, Good see *Chenopodium bonushenricus* pl. 72
Kingcup see *Caltha palustris* pl. 4
Knapweed, Great see *Centaurea scabiosa* pl. 49

Lady's Mantle, Alpine see *Alchemilla alpina* pl. 26
Lady's Slipper see *Cypripedium calceolus* pl. 82
Lady's Smock see *Cardamine pratensis* pl. 7
Lady's Tresses, Creeping see *Goodyera repens* pl. 80
Large Bitter Cress see *Cardamine amara* pl. 7
Large Lizard Clover see *Trifolium molinerii* pl. 22
Large Wintergreen see *Pyrola rotundifolia* pl. 56
Least Yellow Trefoil see *Trifolium micranthum* pl. 23
Leek, Sand see *Allium scorodoprasum* pl. 84
Lesser Bugloss see *Lycopsis arvensis* pl. 60
Lesser Butterfly Orchid see *Platanthera bifolia* pl. 82
Lesser Calamint see *Calamintha nepeta* pl. 68
Lesser Celandine see *Ranunculus ficaria* pl. 3
Lesser Centaury see *Centaurium pulchellum* pl. 58
Lesser Fleabane see *Pulicaria vulgaris* pl. 45
Lesser Valerian see *Valeriana dioica* pl. 43
Lesser Water Forget-me-not see *Myosotis caespitosa* pl. 60
Lettuce, Alpine see *Cicerbita alpina* pl. 53
Lily, May see *Maiathemum bifolium* pl. 84
Lily of the Valley see *Convallaria majalis* pl. 84
Lily, Pyrenean see *Lilium pyrenaicum* pl. 85
Lime, Small-leaved see *Tilia cordata* pl. 18
Liquorice, Wild see *Astralagus glycyphyllos* pl. 24
Little Three-lobed Crowfoot see *Ranunculus tripartitus* pl. 2
Little Tree Willow see *Salix arbuscula* pl. 78
Lizard Clover, Large see *Trifolium molinerii* pl. 22
Lizard Orchid see *Himantoglossum hircinum* pl. 81
Lobed Poplar see *Populus alba* pl. 78
Lobelia, Blue see *Lobelia urens* pl. 54
Loddon Pondweed see *Potamogeton nodosus* pl. 89
London Pride see *Saxifraga* × *urbium* pl. 32
Long Smooth-headed Poppy see *Papaver dubium* pl. 5
Long-leaved Dock see *Rumex longifolius* pl. 74
Long-leaved Helleborine see *Cephalanthera longifolia* pl. 80
Long-leaved Sundew see *Drosera intermedia* pl. 34
Loosestrife, Purple see *Lythrum salicaria* pl. 34
Loosestrife, Yellow see *Lysimachia vulgaris* pl. 57
Lovage see *Ligusticum scoticum* pl. 40
Lucerne see *Medicago sativa* pl. 21

Madder see *Rubia peregrina* pl. 42
Maiden Pink see *Dianthus deltoides* pl. 13
Mallow, Hispid Marsh see *Althaea hirsuta* pl. 18
Mallow, Marsh see *Althaea officinalis* pl. 18
Mallow, Musk see *Malva moschata* pl. 18
Mallow, Tree see *Lavatera arborea* pl. 18
Man Orchid see *Aceras anthropophorum* pl. 81, pl. 82
Many-seeded Goose-foot see *Chenopodium polyspermum* pl. 71
Mare's Tail see *Hippuris vulgaris* pl. 34
Marigold, Bur see *Bidens tripartita* pl. 45
Marigold, Marsh see *Caltha palustris* pl. 4
Marsh Bird's-foot Trefoil see *Lotus uliginosus* pl. 23
Marsh Cinquefoil see *Potentilla palustris* pl. 27
Marsh Forget-me-not see *Myosotis secunda* pl. 60
Marsh Gentian see *Gentiana pneumonanthe* pl. 59

Marsh Helleborine see *Epipactis palustris* pl. 81
Marsh Mallow see *Althaea officinalis* pl. 18
Marsh Mallow, Hispid see *Althaea hirsuta* pl. 18
Marsh Marigold see *Caltha palustris* pl. 4
Marsh Thistle see *Cirsium palustra* pl. 48
Marsh Woundwort see *Stachys palustris* pl. 69
Marsh Yellow Cress see *Rorippa palustris* pl. 7
Marsh-wort, Procumbent see *Apium nodiflorum* pl. 37
May Lily see *Maiathemum bifolium* pl. 84
Mayweed, Rayless see *Matricaria matricarioides* pl. 46
Meadow Clary see *Salvia pratensis* pl. 68
Meadow Rue, Cliff see *Thalictrum minus* pl. 1
Meadow Saxifrage see *Saxifraga granulata* pl. 32
Meadow Sweet see *Filipendula ulmaria* pl. 26
Melilot, Yellow see *Melilotus altissima* pl. 22
Mignonette, Wild see *Reseda lutea* pl. 11
Milfoil, Whorled Water see *Myriophyllum verticillatum* pl. 34
Milk-vetch see *Astragalus glycyphyllos* pl. 24
Milk-vetch, Alpine see *Astragalus alpinus* pl. 24
Mint, Cat see *Nepeta cataria* pl. 68
Mint, Pepper see *Mentha × piperita* pl. 67
Mint, Round-leaved see *Mentha suaveolens* pl. 67
Mollyblobs see *Caltha palustris* pl. 4
Mossy Saxifrage see *Saxifraga hypnoides* pl. 32
Mountain Ash see *Sorbus aucuparia* pl. 31
Mountain Avens see *Dryas octopetala* pl. 29
Mountain Rock-Cress see *Cardaminopsis petraea* pl. 7
Mud Rush, Salt see *Juncus gerardii* pl. 86
Mugwort see *Artemisia vulgaris* pl. 46
Mullein, Common see *Verbascum thapsus* pl. 62
Mullein, Dark see *Verbascum nigrum* pl. 62
Musk Mallow see *Malva moschata* pl. 18
Musky Storksbill see *Erodium moschatum* pl. 20
Mustard, Treacle see *Erysimum cheiranthoides* pl. 9
Myrtle-leaved Willow see *Salix myrsinites* pl. 78

Narrow-leaved Bird's-foot Trefoil see *Lotus tenuis* pl. 23
Narrow-leaved Bitter Cress see *Cardamine impatiens* pl. 8
Narrow-leaved St John's Wort see *Hypericum linarifolium* pl. 17
Narrow-leaved Vetch, see *Vicia angustifolia* pl. 25
Narrow-lipped Helleborine see *Epipactis leptochila* pl. 80
Nettle, Stinging see *Urtica dioica* pl. 76
Night-flowering Catchfly see *Silene noctiflora* pl. 14
Nightshade, Alpine Enchanter's see *Circaea alpina* pl. 35
Nightshade, Black see *Solanum nigrum* pl. 61
Nightshade, Deadly see *Atropa belladonna* pl. 61
Nightshade, Woody see *Solanum dulcamara* pl. 61
Nipple-wort see *Lapsana communis* pl. 50
Nodding Catchfly see *Silene nutans* pl. 13
Nodding Thistle see *Carduus nutans* pl. 48
Northern Bedstraw see *Galium boreale* pl. 42
Northern Shore-wort see *Mertensia maritima* pl. 60
Notthingham Catchfly see *Silene nutans* pl. 13

Old Man's Beard see *Clematis vitalba* pl. 1
One-flowered Wintergreen see *Moneses uniflora* pl. 56
Opium Poppy see *Papaver somniferum* pl. 5
Opposite-leaved Pondweed see *Groenlandia densa* pl. 89
Orache, Common see *Atriplex patula* pl. 72
Orache, Grass-leaved see *Atriplex littoralis* pl. 72
Orache, Hastate see *Atriplex hastata* pl. 72
Orchid, Fen see *Liparis loeselii* pl. 80
Orchid, Frog see *Coeloglossum viride* pl. 82
Orchid, Lesser Butterfly see *Platanthera bifolia* pl. 82
Orchid, Lizard see *Himantoglossum hircinum* pl. 81
Orchid, Man see *Aceras anthropophorum* pl. 81, pl. 82
Orchid, Small White see *Leucorchis albida* pl. 82
Orchid, Sweet-scented see *Gymnadenia conopsea* pl. 82
Ox-tongue, Hawkweed see *Picris hieracioides* pl. 50
Oxlip see *Primula elatior* pl. 57

Paeony see *Paeonia mascula* pl. 4
Pale Heath Violet see *Viola lactea* pl. 12
Pale Persicaria see *Polygonum lapathifolium* pl. 73
Pansy, Tricolor see *Viola tricolor* pl. 12
Parsley Piert see *Aphanes arvensis* pl. 26
Parsley, Spreading Hedge see *Torilis arvensis* pl. 40
Parsnip, Cow see *Heracleum sphondylium* pl. 40
Parsnip, Water see *Sium latifolium* pl. 37
Pasque Flower see *Pulsatilla vulgaris* pl. 1
Pea, Wild see *Lathyrus sylvestris* pl. 25
Pear see *Pyrus pyraster* pl. 31
Pearlwort, Common see *Sagina procumbens* pl. 16
Pedicelled Willow Herb see *Epilobium roseum* pl. 35
Pellitory of the Wall see *Parietaria judaica* pl. 76

Penny Cress, Perfoliate see *Thlaspi perfoliatum* pl. 10
Pennywort see *Umbilicus rupestris* pl. 33
Pepper Mint see *Mentha × piperita* pl. 67
Pepper, Water see *Polygonum hydropiper* pl. 73
Perennial Centaury see *Centaurium scilloides* pl. 58
Perennial Sea Spurrey see *Spergularia media* pl. 16
Perfoliate Penny Cress see *Thlaspi perfoliatum* pl. 10
Perfoliate Pondweed see *Potamogeton perfoliatus* pl. 89
Perforate St John's Wort see *Hypericum perforatum* pl. 17
Persian Speedwell see *Veronica persica* pl. 63
Persicaria, Amphibious see *Polygonum amphibium* pl. 73
Persicaria, Common see *Polygonum persicaria* pl. 73
Persicaria, Pale see *Polygonum lapathifolium* pl. 73
Persicaria, Spotted see *Polygonum nodosum* pl. 73
Piert, Parsley see *Aphanes arvensis* pl. 26
Pimpernel, Yellow see *Lysimachia nemorum* pl. 57
Pink, Maiden see *Dianthus deltoides* pl. 13
Plum, Wild see *Prunus domestica* pl. 26
Pondweed, Canadian see *Elodea canadensis* pl. 79
Pondweed, Grassy see *Potamogeton obtusifolius* pl. 90
Pondweed, Loddon see *Potamogeton nodosus* pl. 89
Pondweed, Opposite-leaved see *Groenlandia densa* pl. 89
Pondweed, Perfoliate see *Potamogeton perfoliatus* pl. 89
Pondweed, Small see *Potamogeton berchtoldii* pl. 90
Pondweed, Various-leaved see *Potamogeton gramineus* pl. 89
Poplar, Grey see *Populus canescens* pl. 78
Poplar, Lobed see *Populus alba* pl. 78
Poplar, White see *Populus alba* pl. 78
Poppy, Long Smooth-headed see *Papaver dubium* pl. 5
Poppy, Opium see *Papaver somniferum* pl. 5
Poppy, Welsh see *Meconopsis cambrica* pl. 5
Pride, London see *Saxifraga × urbium* pl. 32
Primrose see *Primula vulgaris* pl. 57
Primrose, Bird's-eye see *Primula farinosa* pl. 57
Procumbent Marsh-wort see *Apium nodiflorum* pl. 37
Purple Dead-nettle see *Lamium Purpureum* pl. 70
Purple Loosestrife see *Lythrum salicaria* pl. 34
Purple Saxifrage see *Saxifraga oppositifolia* pl. 32
Purple Spurge see *Euphorbia peplis* pl. 75
Purple Toadflax see *Linaria purpurea* pl. 62
Purse, Shepherd's see *Capsella bursa-pastoris* pl. 10
Purslane, Sea see *Honkenya peploides* pl. 16
Pyrenean Lily see *Lilium pyrenaicum* pl. 85

Radish, Sea see *Raphanistrum maritimus* pl. 11
Ragged Robin see *Lychnis flos-cuculi* pl. 14
Ragwort, Hoary see *Senecio erucifolius* pl. 47
Raspberry see *Rubus idaeus* pl. 28
Rattle, Red see *Pedicularis palustris* pl. 65
Rayless Mayweed see *Matricaria matricarioides* pl. 46
Red Campion see *Silene dioica* pl. 14
Red Currant, Upright see *Ribes spicatum* pl. 33
Red Goose-foot see *Chenopodium rubrum* pl. 72
Red Rattle see *Pedicularis palustris* pl. 65
Restharrow see *Ononis repens* pl. 21
River Crowfoot see *Ranunculus fluitans* pl. 2
Robin, Ragged see *Lychnis flos-cuculi* pl. 14
Rock Cinquefoil see *Potentilla rupestris* pl. 27
Rock Spurrey see *Spergularia rupicola* pl. 16
Rock Stonecrop see *Sedum forsteranum* pl. 33
Rock-Cress, Mountain see *Cardaminopsis petraea* pl. 7
Rocket, Common Yellow see *Barbarea vulgaris* pl. 7
Rocket, Dyer's see *Reseda luteola* pl. 11
Rocket, Sea see *Cakile maritima* pl. 11
Rockrose, Common see *Helianthemum nummularium* pl. 11
Rockrose, Hoary see *Helianthemum canum* pl. 11
Rockrose, White see *Helianthemum apenninum* pl. 11
Rod, Golden see *Solidago virgaurea* pl. 44
Root, Coral see *Cardamine bulbifera* pl. 8
Rose Bay see *Epilobium angustifolium* pl. 35
Rose, Dog see *Rosa canina* pl. 30
Rose, Downy see *Rosa tomentosa* pl. 30
Rose, Guelder see *Viburnum opulus* pl. 41
Rose of Sharon see *Hypericum calycinum* pl. 17
Rose, Soft-leaved see *Rosa villosa* pl. 30
Roseroot see *Rhodiola rosea* pl. 33
Rough Hawkbit see *Leontodon hispidus* pl. 53
Rough Hawk's-beard see *Crepis biennis* pl. 50
Round-fruited Rush see *Juncus compressus* pl. 86
Round-leaved Mint see *Mentha suaveolens* pl. 67
Rowan see *Sorbus aucuparia* pl. 31
Rue, Cliff Meadow see *Thalictrum minus* pl. 1
Rush, Capitate see *Juncus capitatus* pl. 87
Rush, Chestnut see *Juncus castaneus* pl. 87
Rush, Common see *Juncus subuliflorus* pl. 86

Rush, Flowering see *Butomus umbellatus* pl. 79
Rush, Round-fruited see *Juncus compressus* pl. 86
Rush, Salt Mud see *Juncus gerardii* pl. 86
Rush, Sea see *Juncus maritimus* pl. 87
Rush, Sharp Sea see *Juncus acutus* pl. 87
Rush, Soft see *Juncus effusus* pl. 86

Sage, Wood see *Teucrium scorodonia* pl. 70
Sainfoin see *Onobrychis viciifolia* pl. 24
St John's Wort, Narrow-leaved see *Hypericum linarifolium* pl. 17
St John's Wort, Wavy see *Hypericum undulatum* pl. 17
Sallow, Great see *Salix caprea* pl. 78
Salt Mud Rush see *Juncus gerardii* pl. 86
Sand Leek see *Allium scorodoprasum* pl. 84
Sandwort, Arctic see *Arenaria norvegica* pl. 15
Sandwort, Irish see *Arenaria ciliata* pl. 15
Sandwort, Three-veined see *Moehringia trinervia* pl. 15
Saw-wort see *Serratula tinctoria* pl. 49
Saxifrage, Bulbous see *Saxifraga granulata* pl. 32
Saxifrage, Burnet see *Pimpinella saxifraga* pl. 38
Saxifrage, Meadow see *Saxifraga granulata* pl. 32
Saxifrage, Mossy see *Saxifraga hypnoides* pl. 32
Saxifrage, Purple see *Saxifraga oppositifolia* pl. 32
Saxifrage, Tufted see *Saxifraga cespitosa* pl. 32
Scentless Chamomile see *Tripleurospermum martitimum* pl. 46
Scottish Thistle see *Onopordon acanthium* pl. 49
Sea Aster see *Aster tripolium* pl. 44
Sea Beet see *Beta vulgaris* pl. 72
Sea Holly see *Eryngium maritimum* pl. 36
Sea Kale see *Crambe maritima* pl. 11
Sea Purslane see *Honkenya peploides* pl. 16
Sea Radish see *Raphanistrum maritimus* pl. 11
Sea Rocket see *Cakile maritima* pl. 11
Sea Rush see *Juncus maritimus* pl. 87
Sea Rush, Sharp see *Juncus acutus* pl. 87
Sea Spurge see *Euphorbia paralias* pl. 75
Sea Spurrey, Perennial see *Spergularia media* pl. 16
Sea Stock see *Matthiola sinuata* pl. 6
Sea Wormwood see *Artemisia maritima* pl. 46
Seaside Crowfoot see *Ranunculus baudotii* pl. 2
Sedge, Bladder see *Carex vesicaria* pl. 94
Self-heal see *Prunella vulgaris* pl. 69
Service Tree, Wild see *Sorbus torminalis* pl. 31
Sharp Dock see *Rumex conglomeratus* pl. 74
Sharp Sea Rush see *Juncus acutus* pl. 87
Sheep's Sorrel see *Rumex acetosella* pl. 74
Shepherd's Purse see *Capsella bursa-pastoris* pl. 10
Shore Dock see *Rumex rupestris* pl. 74
Shore-wort, Northern see *Mertensia maritima* pl. 60
Short-haired Eyebright see *Euphrasia brevipila* pl. 64
Short-leaved Forget-me-not see *Myosotis stolonifera* pl. 60
Shrubby Speedwell see *Veronica fruticans* pl. 64
Silver Birch see *Betula pendula* pl. 76
Silverweed see *Potentilla anserina* pl. 27
Skull-cap see *Scutellaria galericulata* pl. 68
Slender-Flowered Thistle see *Carduus tenuiflorus* pl. 48
Slipper, Lady's see *Cypripedium calceolus* pl. 82
Sloe see *Prunus spinosa* pl. 26
Small Bur-reed see *Sparganium minimum* pl. 88
Small Furze see *Ulex minor* pl. 21
Small Pondweed see *Potamogeton berchtoldii* pl. 90
Small Teasel see *Dipsacus pilosus* pl. 43
Small White Orchid see *Leucorchis albida* pl. 82
Small-flowered Cranesbill see *Geranium pusillum* pl. 19
Small-leaved Lime see *Tilia cordata* pl. 18
Smock, Lady's see *Cardamine pratensis* pl. 7
Smooth Hawkbit see *Leontodon autumnalis* pl. 53
Smooth-headed Poppy, Long see *Papaver dubium* pl. 5
Snakeweed see *Polygonum bistorta* pl. 73
Soapwort see *Saponaria officinalis* pl. 13
Soft Hawk's-beard see *Crepis mollis* pl. 50
Soft Rush see *Juncus effusus* pl. 86
Soft-leaved Rose see *Rosa villosa* pl. 30
Soldier, Water see *Stratiotes aloides* pl. 79
Sorrel, Common see *Rumex acetosa* pl. 74
Sorrel, Sheep's see *Rumex acetosella* pl. 74
Sour Cherry see *Prunus cerasus* pl. 26
Sowthistle, Corn see *Sonchus arvensis* pl. 53
Sowthistle, Fen see *Sonchus palustris* pl. 53
Spanish Catchfly see *Silene otites* pl. 13
Spathulate Fleawort see *Senecio spathulifolius* pl. 47
Spear Thistle see *Cirsium vulgare* pl. 48
Spearwort, Great see *Ranunculus reptans* pl. 3
Speedwell, Grey see *Veronica polita* pl. 63
Speedwell, Persian see *Veronica persica* pl. 63

Speedwell, Shrubby see *Veronica fruticans* pl. 64
Speedwell, Spiked see *Veronica × spicata* pl. 64
Speedwell, Spring see *Veronica verna* pl. 63
Spiked Speedwell see *Veronica spicata* pl. 64
Spindle see *Euonymus europaeus* pl. 20
Spotted Persicaria see *Polygonum nodosum* pl. 73
Spreading Bell-flower see *Campanula patula* pl. 54
Spreading Hedge Parsley see *Torilis arvensis* pl. 40
Spring Speedwell see *Veronica verna* pl. 63
Spurge, Broad-leaved see *Euphorbia platyphyllos* pl. 75
Spurge, Irish see *Euphorbia hyberna* pl. 75
Spurge, Purple see *Euphorbia peplis* pl. 75
Spurge, Sea see *Euphorbia paralias* pl. 75
Spurge, Sun see *Euphorbia helioscopia* pl. 75
Spurrey, Perennial Sea see *Spergularia media* pl. 16
Spurrey, Rock see *Spergularia rupicola* pl. 16
Star Thistle see *Centaurea calcitrapa* pl. 49
Stinging Nettle see *Urtica dioica* pl. 76
Stinking Hawk's-beard see *Crepis foetida* pl. 50
Stinking Iris see *Iris foetidissima* pl. 83
Stitchwort, Greater see *Stellaria holostea* pl. 15
Stock, Sea see *Matthiola sinuata* pl. 6
Stonecrop, Rock see *Sedum forsteranum* pl. 33
Storksbill, Common see *Erodium cicutarium* pl. 20
Storksbill, Musky see *Erodium moschatum* pl. 20
Strap-wort see *Corrigiola litoralis* pl. 71
Strawberry, Barren see *Potentilla sterilis* pl. 27
Strawberry Tree see *Arbutus unedo* pl. 55
Strawberry, Wild see *Fragaria vesca* pl. 27
Strawberry-headed Clover see *Trifolium fragiferum* pl. 23
Striated Catchfly see *Silene conica* pl. 13
Sun Spurge see *Euphorbia helioscopia* pl. 75
Sundew see *Drosera rotundifolia* pl. 34
Sundew, Long-leaved see *Drosera intermedia* pl. 34
Sweet Cicely see *Myrrhis odorata* pl. 38
Sweet Galingale see *Cyperus longus* pl. 91
Sweet, Meadow see *Filipendula ulmaria* pl. 26
Sweet Violet see *Viola odorata* pl. 12
Sweet Woodruff see *Galium odoratum* pl. 42
Sweet-scented Orchid see *Gymnadenia conopsea* pl. 82

Tall Broomrape see *Orobanche elatior* pl. 66
Tare, Hairy see *Vicia hirsuta* pl. 24
Teasel, Common see *Dipsacus fullonum* pl. 43
Teasel, Small see *Dipsacus pilosus* pl. 43
Thistle, Ground see *Cirsium acaule* pl. 48
Thistle, Marsh see *Cirsium palustre* pl. 48
Thistle, Nodding see *Carduus nutans* pl. 48
Thistle, Scottish see *Onopordon acanthium* pl. 49
Thistle, Slender-flowered see *Carduus tenuiflorus* pl. 48
Thistle, Spear see *Cirsium vulgare* pl. 48
Thistle, Star see *Centaurea calcitrapa* pl. 49
Three-lobed Crowfoot, Little see *Ranunculus tripartitus* pl. 2
Three-veined Sandwort see *Moehringia trinervia* pl. 15
Toadflax, Common see *Linaria vulgaris* pl. 62
Toadflax, Purple see *Linaria purpurea* pl. 62
Tongue, Green Hound's see *Cynoglossum germanicum* pl. 59
Tongue, Hound's see *Cynoglossum officinale* pl. 59
Tormentil, Common see *Potentilla erecta* pl. 27
Touch-me-not see *Impatiens nolitangere* pl. 20
Tower-Cress, Glabrous see *Arabis glabra* pl. 7
Traveller's Joy see *Clematis vitalba* pl. 1
Treacle Mustard see *Erysimum cheiranthoides* pl. 9
Tree Mallow, Lesser see *Lavatera cretica* pl. 18
Tree Willow, Little see *Salix arbuscula* pl. 78
Trefoil, Bird's-foot see *Lotus corniculatus* pl. 23
Trefoil, Hop see *Trifolium campestre* pl. 23
Trefoil, Least Yellow see *Trifolium micranthum* pl. 23
Trefoil, Marsh Bird's-foot see *Lotus uliginosus* pl. 23
Trefoil, Narrow-leaved Bird's-foot see *Lotus tenuis* pl. 23
Tresses, Creeping Lady's see *Goodyera repens* pl. 80
Triangular-stalked Garlic see *Allium triquetrum* pl. 85
Tricolor Pansy see *Viola tricolor* pl. 12
Tuberous Caraway see *Bunium bulbocastanum* pl. 37
Tuberous Comfrey see *Symphytum tuberosum* pl. 60
Tufted Saxifrage see *Saxifraga cespitosa* pl. 32
Tulip, Wild see *Tulipa sylvestris* pl. 85

Unbranched Bur-reed see *Sparganium emersum* pl. 88
Upright Red Currant see *Ribes spicatum* pl. 33

Valerian see *Valeriana officinalis* pl. 43
Valerian, Lesser see *Valeriana dioica* pl. 43
Various-leaved Pondweed see *Potamogeton gramineus* pl. 89

Venus' Looking-glass see *Legousia hybrida* pl. 54
Vetch, Kidney see *Anthyllis vulneraria* pl. 23
Vetch, Narrow-leaved see *Vicia angustifolia* pl. 25
Vetch, Wood Bitter see *Vicia orobus* pl. 24
Vetch, Yellow see *Vicia lutea* pl. 25
Vetchling, Yellow see *Lathyrus aphaca* pl. 25
Violet, Dame's see *Hesperis matronalis* pl. 8
Violet, Pale Heath see *Viola lactea* pl. 12
Violet, Sweet see *Viola odorata* pl. 12
Violet, Woodland see *Viola reichenbachiana* pl. 12
Viper's Bugloss see *Echium vulgare* pl. 61

Warren Crocus see *Romulea columnae* pl. 83
Water Avens see *Geum rivale* pl. 29
Water Cress see *Nasturtium officinale* pl. 7
Water Dropwort, Corky-fruited see *Oenanthe pimpinelloides* pl. 39
Water Dropwort, Hemlock see *Oenanthe crocata* pl. 39
Water Figwort see *Scrophularia auriculata* pl. 63
Water Forget-me-not, Lesser see *Myosotis caespitosa* pl. 60
Water Germander see *Teucrium scordium* pl. 70
Water Milfoil, Whorled see *Myriophyllum verticillatum* pl. 34
Water Parsnip see *Sium latifolium* pl. 37
Water Pepper see *Polygonum hydropiper* pl. 73
Water Soldier see *Stratiotes aloides* pl. 79
Wavy St John's Wort see *Hypericum undulatum* pl. 17
Weld see *Reseda luteola* pl. 11
Welsh Poppy see *Meconopsis cambrica* pl. 5
White Beam see *Sorbus aria* pl. 31
White Beam, Broad-leaved see *Sorbus latifolia* pl. 31
White Bryony see *Bryonia dioica* pl. 36
White Campion see *Silene alba* pl. 14
White Dead-nettle see *Lamium album* pl. 70
White Helleborine see *Cephalanthera damasonium* pl. 80
White Orchid, Small see *Leucorchis albida* pl. 82
White Poplar see *Populus alba* pl. 78
White Rockrose see *Helianthemum apenninum* pl. 11
Whorled Water Milfoil see *Myriophyllum verticillatum* pl. 34
Whortleberry see *Vaccinium myrtillus* pl. 55
Wild Angelica see *Angelica sylvestris* pl. 40
Wild Asparagus see *Asparagus officinalis* pl. 84
Wild Balsam see *Impatiens nolitangere* pl. 20
Wild Cabbage see *Brassica oleracea* pl. 9
Wild Cherry see *Prunus avium* pl. 26
Wild Liquorice see *Astragalus glycyphyllos* pl. 24
Wild Mignonette see *Reseda lutea* pl. 11
Wild Pea see *Lathyrus sylvestris* pl. 25
Wild Plum see *Prunus domestica* pl. 26
Wild Service Tree see *Sorbus torminalis* pl. 31
Wild Strawberry see *Fragaria vesca* pl. 27
Wild Tulip see *Tulipa sylvestris* pl. 85
Willow, Almond see *Salix triandra* pl. 77
Willow, Crack see *Salix fragilis* pl. 77
Willow, Dark-leaved see *Salix nigricans* pl. 78
Willow, Downy see *Salix lapponum* pl. 77
Willow Herb, Alpine see *Epilobium anagallidifolium* pl. 35
Willow Herb, Great see *Epilobium hirsutum* pl. 35
Willow Herb, Hairy see *Epilobium parviflorum* pl. 35
Willow Herb, Pedicelled see *Epilobium roseum* pl. 35
Willow, Little Tree see *Salix arbuscula* pl. 78
Willow, Myrtle-leaved see *Salix myrsinites* pl. 78
Winter Cress see *Barbarea vulgaris* pl. 7
Wintergreen, Large see *Pyrola rotundifolia* pl. 56
Wintergreen, One-flowered see *Moneses uniflora* pl. 56
Wood, Anemone see *Anemone nemorosa* pl. 1
Wood Bitter Vetch see *Vicia orobus* pl. 24
Wood Club-rush see *Scirpus sylvaticus* pl. 91
Wood Cranesbill see *Geranium sylvaticum* pl. 19
Wood Groundsel see *Senecio sylvaticus* pl. 47
Wood Sage see *Teucrium scorodonia* pl. 70
Woodland Violet see *Viola reichenbachiana* pl. 12
Woodruff, Sweet see *Galium odoratum* pl. 42
Wood-rush, Forster's see *Luzula forsteri* pl. 87
Wood-rush, Great see *Luzula sylvatica* pl. 87
Wood-rush, Hairy see *Luzula pilosa* pl. 87
Woody Nightshade see *Solanum dulcamara* pl. 61
Wormwood see *Artemisia absinthium* pl. 46
Wormwood, Breckland see *Artemisia campestris* pl. 46
Wormwood, Sea see *Artemisia maritima* pl. 46
Woundwort, Alpine see *Stachys alpina* pl. 69
Woundwort, Marsh see *Stachys palustris* pl. 69
Wych Elm see *Ulmus glabra* pl. 76

Yellow Bartsia see *Parentucellia viscosa* pl. 65
Yellow Corydalis see *Corydalis lutea* pl. 6
Yellow Cress, Amphibious see *Rorippa amphibia* pl. 7

Yellow Cress, Creeping see *Rorippa sylvestris* pl. 7
Yellow Cress, Marsh see *Rorippa palustris* pl. 7
Yellow Horned-poppy see *Glaucium flavum* pl. 5
Yellow Loosestrife see *Lysimachia vulgaris* pl. 57
Yellow Melilot see *Melilotus altissima* pl. 22
Yellow Pimpernel see *Lysimachia nemorum* pl. 57
Yellow Rocket, Common see *Barbarea vulgaris* pl. 7
Yellow Trefoil, Least see *Trifoliur micranthum* pl. 23
Yellow Vetch see *Vicia lutea* pl. 2
Yellow Vetchling see *Lathyri aphaca* pl. 25